591

BIRD *Flight*

BIRD flight

AN ILLUSTRATED STUDY OF BIRDS' AERIAL MASTERY

ROBERT BURTON

FOREWORD BY

DR CHRISTOPHER PERRINS
DIRECTOR, EDWARD GREY INSTITUTE OF FIELD ORNITHOLOGY
UNIVERSITY OF OXFORD

Facts On File
New York • Oxford • Sydney

Facts On File, Inc.
460 Park Avenue South
New York NY 10016
USA

Facts On File Limited
Collins Street
Oxford OX4 1XJ
United Kingdom

Facts On File Pty Ltd
Talavera & Khartoum Rds
North Ryde NSW 2113
Australia

Library of Congress Cataloging-in-Publication Data
Burton, Robert, 1941-
    Birdflight/Robert Burton
       p.   cm.
    Includes bibliographical references.
    ISBN 0–8160–2410–3 (alk. paper)
    1. Birds—Flight.   I. Title
    QL698.7.B87 1990
    598.2'1852—dc20   89–25970
                                       CIP

ISBN 0–8160–2410–3

A British CIP catalogue record for this book is available from the British Library.

Australian CIP data available on request from Facts On File.

Facts On File books are available at special discounts when purchased in bulk quantities for businesses, associations, institutions or sales promotions. Please call our Special Sales Department in New York at 212/683-2244 (dial 800/322-8755 except in NY, AK or HI) or in Oxford at 0865/728399.

10 9 8 7 6 5 4 3 2 1

An Eddison · Sadd Edition
Edited, designed and produced by
Eddison Sadd Editions Limited
St Chad's Court, 146B King's Cross Road
London WC1X 9DH

Phototypeset by Bookworm Typesetting, Manchester, England
Origination by Columbia Offset, Singapore
Printed and bound by Clays Ltd, Bungay, Suffolk, England

FRONTISPIECE
An Andean condor soars on outstretched wings.

CONTENTS

FOREWORD 6

INTRODUCTION 8

Chapter 1 CONQUERING THE AIR 12
Into the air 14 ● Evolution of bird flight 15 ● Learning how birds fly 19

Chapter 2 THE BIRD AS A FLYING MACHINE 22
Basic principles 24 ● The wing of a bird 28 ● Spreading the wing 30 ● Feathers 30
Gliding 32 ● The tail 37 ● Flapping flight 38 ● Power for flight 42 ● Power and speed 43
What speed to fly 44 ● Built for flight 50 ● Fuel for flying 53 ● Heart and lungs 54
Flight maintenance 55 ● Moulting 58

Chapter 3 FLYING SKILLS 60
Taking off 62 ● Landing 70 ● Control of flight 77 ● Fine control 83

Chapter 4 STYLES OF FLIGHT 88
Wings and flight 90 ● Free-wheeling 96 ● Soaring 98 ● Ground effect 108
Formation flying 111 ● Hovering 113

Chapter 5 LIVING IN THE AIR 118
Time in the air 120 ● Flight for feeding 121 ● Flying to avoid predators 132
Flight for migration 135 ● Flight for breeding 144 ● First flight 150

LIST OF SCIENTIFIC NAMES 155
FURTHER READING 157
INDEX 158
ACKNOWLEDGEMENTS 160

FOREWORD

The ability to fly is surely the most remarkable feature of birds. Other creatures such as insects and bats can fly, but it is birds that have made the sky their own.

Nevertheless, for all the work that has been done on aerodynamics, we still know remarkably little about some aspects of bird flight. Aircraft engineers have tended to study the fixed wings of aeroplanes. But the bird's wing is flexible; birds can alter not only the size of the wing, but also its shape. Look, for example, at the tips of the flight feathers of the owl on page 94; they not only bend, but also twist, changing in shape from moment to moment during the wing beat. Small wonder then that we find many of the details of bird flight hard to understand.

Much of what we do know about the flight of birds is tucked away in the scientific literature. Robert Burton has done a fine job of making this literature available – in jargon-free form – to the general reader. He has made it easy for us to pick up the essentials in a way which enables us to understand not only how birds fly, but also how the wide variety of wing shapes serve the differing needs of the different species. Even if you choose to avoid the text – and you shouldn't – the striking collection of photographs of birds in flight is something one can look through again and again.

Dr Christopher Perrins
Director, Edward Grey Institute of Field Ornithology,
University of Oxford

African spoonbills fly leisurely over Lake Turkana in Kenya. Their broad wings demonstrate a flexibility of structure that contrasts with the fixed wings of aircraft, and gives birds an efficiency in flight that is impossible in man-made machines.

INTRODUCTION

THE FASCINATION OF BIRD FLIGHT

Early June in the High Arctic is springtime. The ground is still snow-covered but the birds have returned from southern countries where they had fled to escape the darkness and intense cold of the polar winter. They have flown thousands of kilometres and are preparing for the nesting season. It is the third time that I have come to the Arctic to watch the arrival of the birds. The most exciting has always been the knot, a species of wader which winters on European coasts as a rather dirty grey bird but is transformed by its spring moult into a handsome chestnut and cinnamon. Even more delightful than the summer plumage is the song-flight of the male. As each one appears on the tundra, he stakes out a territory and advertises loudly and persistently for a mate.

There is 24-hour daylight in the Arctic at this time of the year and the male knots never cease their activity. They fly up to a height of 100 to 200 metres (330 to 660 feet), where they are little more than dots in the sky, then fly in circles. Long glides, during which the knot floats with outstretched wings and fanned tail pale against the blue sky, alternate with bursts of rapid wingbeats as the bird regains height lost in the glide, and a haunting *poorr-mee*, *poorr-poorr*, reminiscent of the curlew's call, floats down to earthbound listeners and announces that life is returning to the tundra.

I have tried following individual knots on their song-flights but it is all too easy to lose them as they circle up and down the valley and disappear against the mountain backdrop. Some definitely stay up for over an hour and some probably continue the song-flight for much longer. During this period the male knots have little time to feed; they are either off on their song-flights or chasing intruding males. Respite comes only when they acquire mates. They continue to chase trespassing males out of their territories but the song-flights cease abruptly when a female settles with them.

Apart from the spectacle of the song-flight, the behaviour of the male knot at this time is remarkable because it takes place immediately after an epic flight of thousands of kilometres. Our knots had come from the coasts of Western Europe, with a stop in Iceland or Norway. This is a long way for a small bird, but the knots arrive with sufficient reserves of fat to spend the next week or so flying in circles and chasing rivals with hardly a break to feed.

This set me thinking about bird flight. We tend to take for granted the fact that birds fly. It is the essence of being a bird and if the body of a sparrow, for instance, is compared with a mouse, it is clear that the fundamental difference between the bird and the mammal is that the sparrow's body has been almost totally designed for flight. Birds are unlike any other animal, although their scaly legs hint at a reptilian ancestry and there is a school of thought that considers birds to be the last living representatives of the dinosaurs. Yet birds are clearly very different animals from the popular concept of a dinosaur or any other reptile, whereas bats, although also well designed for flight, are still obviously mammals.

Watching the behaviour of the knots highlighted some of the reasons for such a radical evolutionary change in the body of a bird and raised questions about the benefit birds get from their supreme powers of flight. The first conclusion is that feats of endurance, such as long-distance migration, open a new world to knots and other Arctic birds because it gives them the power to take advantage of the short, but very productive, Arctic summer in which to rear their young. Such is the economy of their flight that they arrive with reserves of fat. They can afford to wait for the snow to clear and, meanwhile, they burn their reserves in profligate but very necessary advertisement for a mate. Economy is forgotten in the imperative to breed and contribute to a new generation.

I first became interested in bird flight when I was on a ship bound for the Antarctic and was enchanted when we became the centre of attraction for albatrosses and a host of their smaller relatives in the South Atlantic. I had learned, as a student, of the soaring technique which permits albatrosses to criss-cross the ocean with hardly a wingbeat, but this did not prepare me for the marvel of their skill in exploiting the invisible air currents swirling around the ship.

Knots and albatrosses are very different kinds of birds and the mechanics of their flight are adapted to their particular lifestyle. Just as a knot could not live in the Antarctic waters, neither could an albatross survive on the Arctic tundra. So my interest has been not so much in the way birds fly and their adaptations of anatomy and physiology for overcoming the problems of heavier-than-air flight, as in the way that they use their power of flight to obtain food, find a mate, raise a family, escape enemies and make journeys of migration in search of better living conditions.

It is surprising how little is known about the mechanics of bird flight, considering the popularity of bird-watching. So much has been written about bird life that whole shelves of libraries are allotted to bird books and many scientific journals are devoted entirely to the study of birds, yet there is little about how birds fly. Understanding bird flight – the attribute that underpins nearly every aspect of their biology – has not proved easy. Flight is so difficult to observe. As I write I can

I first became fascinated by the flight skills of albatrosses as they followed my ship bound for the Antarctic. Here a royal albatross hangs in the air with wings partly folded.

see common garden birds flying to and from the bird table. They flash past the window, their wings a blur and their movements hard to discern. Somehow they slow down and land, perhaps upside down to reach some titbit, then take off again and disappear over the hedge, but I cannot make out exactly what they are doing. A camera would freeze their actions and the many photographs on these pages reveal details of wing movement that would not otherwise be seen.

Unfortunately studies of flight have been carried out on only a few species so one of the difficulties in writing a book on bird flight is that the same birds are likely to keep appearing from chapter to chapter. Some are my favourites which I have spent hours watching: albatrosses, flycatchers, waders and geese. Others are the few species which have been studied by professional ornithologists interested in the aerodynamics and physiology of flight: pigeons, gulls, vultures, kestrels, hummingbirds . . . and albatrosses. It will be obvious that I have drawn in detail from other people's studies; my own observations have given me pleasure but are not particularly extensive.

Most of the information in this book comes from the scientific research of zoologists and physicists, especially that of Colin Pennycuik who has been studying bird flight in the laboratory and, more especially, in the wild for over 30 years. Aerodynamics is a difficult and unfamiliar branch of physics which embodies ideas that relate to the complex forces created by the movements of air, an invisible and intangible medium. It is impossible to describe how birds travel through the air without explaining the basic aerodynamics of flight, but I have reduced theory to the point of oversimplification and have used more easily understood analogies of aircraft flight. (Anyone requiring a deeper understanding of the physics of bird flight is advised to consult the books listed on page 156.) Even with simplification, some aspects of the mechanics of bird flight are not easy to grasp, but it is worth the effort because it makes bird-watching so much more enjoyable if you can appreciate how and why the birds are flying.

Bird-flight skills continue to attract my attention nearer home, in fact just outside my study window, where two blue tits compete acrobatically for space at a peanut basket.

CHAPTER 1

CONQUERING THE AIR

How and why birds evolved the ability to fly is not fully understood, but comparisons can be made with present-day gliding animals and we can learn from examining the fossils of the best-known bird ancestor, *Archaeopteryx*.

People have always envied birds their aerial prowess and understanding the principles of flight came from watching birds. But once flight had been mastered, human ingenuity outstripped the pace of evolution and the tables are now turned so that the complexities of bird flight can be examined with the knowledge of aerodynamics gained from aircraft design.

White storks take off at dawn from their perches in a dead tree.

INTO THE AIR

Flight was a remarkable evolutionary breakthrough for the birds because it allowed them to dominate the air with their capacity for fast, agile movement. Although more animals have attempted flight than is generally realized, none have met with the same success as the birds, and many have been extinct for millions of years.

Apart from the birds, the tally of animals that have tried flying includes one species of squid, the gliding frog, three groups of fishes, nine groups of reptiles, seven groups of mammals (including the bats), and nearly all the insects. With the exception of the insects, most of these animals are no more than gliders. They leap into the air, spread some form of wing and rely on gravity to sweep them downwards.

The flying squid takes a running jump, propelling itself up and out of the water, but is little more of a flier than a salmon leaping up a waterfall, and even some octopuses leap out of the water. The flying frog extends its leap by spreading the webs between its toes, and the several gliding lizards and mammals – squirrels, possums and the colugo or flying lemur – glide on wings of skin stretched between elongated ribs or extended limbs. Some of these animals are quite effective gliders. Flying squirrels can glide for 100 metres (330 feet) or more and steer through the tree canopy, or turn round and return to their starting point, but they have no way of improving on a simple glide.

The flying fish that are such a familiar sight to sailors in the tropics have attempted to solve the problem of staying airborne for longer periods. They propel themselves out of the water and glide over the waves, then dip their tail fins into the water, give a few powerful beats, and set off on another glide. Powered flight comes even nearer reality in the hatchet fishes of South America, except that they are not very good at it. Their name derives from the shape of their deep-chested bodies which contain large muscles for working the long pectoral fins. To take off, a hatchet fish taxies along for about 10 metres (33 feet) with its tail beating in the water, then rapidly flaps its fins and rises into the air. The flight does not carry it farther than one or two metres and seems to be a waste of energy – a good leap would be a better way of escaping predators.

Only four groups of animals have truly mastered powered flight. These are the insects, pterosaurs, bats and birds. The insects, uniquely, developed brand-new appendages as wings instead of modifying existing limbs. The pterosaurs had membranes of stretched skin between their limbs, as do the bats, and the birds have modified their forelimbs into wings with the aid of feathers. Unfortunately the fossil record gives little indication of how any of these animals acquired their powers of flight. The gliding animals give a hint of the first stages in the evolution of flight, but there is no 'missing link' that shows how these simple, albeit often efficient, gliders made the major transition to powered, flapping flight. There are no feebly flapping animals, except perhaps the hatchet fishes, and the fossil record is not much help. The earliest fossils of most flying animals are already too far advanced to show the start of a flying ability. However, unique fossils have been discovered that give us some clues about how the power of bird flight evolved.

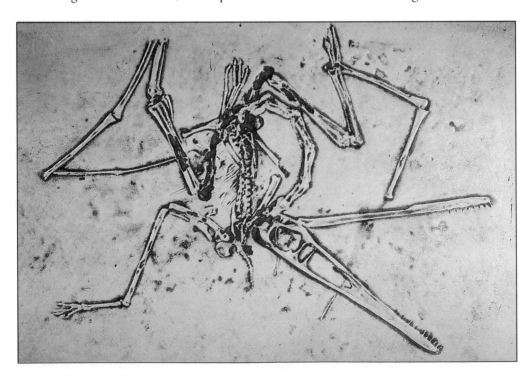

A fossil of the pterosaur Pterodactylus, *showing the well-preserved bones of its long wings. Pterosaurs were successful fliers for millions of years.*

EVOLUTION OF BIRD FLIGHT

The Solenhofen limestone of southern Germany was formed from fine deposits laid down in Jurassic times, 140 million years ago, and compressed into rock that splits into smooth, flat flags that have been quarried for use in lithographic printing. The gentle deposition of sediment onto the bed of an ancient lagoon and the careful splitting of the rock by quarrymen many millions of years later have resulted in the discovery of seven fossils of the earliest known bird, *Archaeopteryx lithographica*. The first fossil to be discovered, in 1861, was no more than a single feather, but the other six fossils are beautifully preserved skeletons of an animal that was part bird and part reptile.

The delicate bones of birds do not fossilize well so our knowledge of the early evolution of bird flight rests entirely on these specimens. If the bodies had not sunk into the water of a sheltered seaside lagoon and been covered with fine material before decay started they would not have been fossilized with their feathers more or less intact and so identified as birds. Even so, two of the fossils were originally described as small dinosaurs called *Composgnathus*, the second being recognized in a Munich museum only in 1988. The excellent state of

BELOW A fossil of Archaeopteryx, *exposed in its bed of limestone, shows clearly the feathers that define it as a bird. Without feathers it is easily mistaken for a small dinosaur.*

FLIGHT FEATHER CLUES

The structure of its flight feathers is good evidence that *Archaeopteryx* could fly rather than merely glide. The narrow leading edge and broad trailing edge of the vane of a flying bird's feather (RIGHT) increases its aerodynamic, lift-producing qualities. The vane of a fossil *Archaeopteryx* wing feather (LEFT) is similarly asymmetrical, whereas the feather of a modern flightless bird is symmetrical (CENTRE).

preservation of these fossils has enabled palaeontologists to piece *Archaeopteryx* together and 'put flesh on the bones' to create an accurate picture of how it looked in life. It has been more difficult to deduce how it lived and how it evolved from a flightless reptilian ancestor. These subjects are still a topic for debate among specialists.

Expert examination of *Archaeopteryx'* bones reveals features of both ancient reptiles and modern birds, of which the following are the most obvious:

Reptile	*Bird*
Teeth	Feathers
Long tail	Forelimb modified into a wing
Simple unjointed ribs	Furcula or wishbone
Claws on 'hands'	Structure of pelvis
Simple brain	Fusion of toe bones
Weak breastbone	Opposable toe (to grip perch)

There has been great discussion in scientific circles about how well *Archaeopteryx* could fly and how it evolved this ability. The form of the feathers is evidence that *Archaeopteryx* was a flying animal. They are very similar to those of a modern bird, not only in the fine detail of the vane and in their shape but also in their arrangement on the bird's body. There are different wing and tail feathers (the remiges and retrices respectively) and the remiges are divided into primary and secondary flight feathers as in a modern bird. The forward-facing part of the vane is narrower than the rearward part, an asymmetry that gives a feather its aerodynamic qualities (see page 36).

The opposable toe on the foot and the claws on its wings suggest that *Archaeopteryx* lived in trees. It would have been able to perch like a modern bird by gripping with its toes, although their claws are not as long and sharp as those of modern perching birds and are more like those of a ground-living chicken. The wing claws probably helped *Archaeopteryx* to clamber through the foliage. A few modern birds, such as the coot and the hoatzin of the Amazonian forests, possess claws on their wings when young and use them when climbing out of the nest.

Although the well-developed wing feathers are a good indication that *Archaeopteryx* could fly, it must have been too heavy and underpowered to be a good flier. None of the fossils has a breastbone and its absence, in specimens which are otherwise so well preserved, suggests that it was probably made of cartilage (like the human breastbone) which would not have fossilized well. This also implies that it did not have the keel which helps to form the anchorage for a modern bird's large flight muscles. This is usually taken to mean that *Archaeopteryx* could not have had well-developed breast muscles, but the bats do not have a keel yet are strong fliers. Their breast muscles are attached to each other and each acts as an anchor for the other. *Archaeopteryx* has a wishbone, which is the two collarbones fused at their tips to make a strut for bracing the shoulders and is a unique feature of modern birds, but it does not have the firm rib cage which in modern birds prevents contraction of the flight muscles squashing the internal organs.

Another theory supporting the idea that *Archaeopteryx* was not very competent in the air was that its brain would not have been able to cope with the complexities of controlling movement in three dimensions, so it needed the long tail to act as a stabilizer like the trailing tail of a child's kite. Recent examination of the fossils has shown, however, that the brain was more complex than originally thought and might have been capable of the fine control of the wings needed to maintain stability without a long tail.

Even if *Archaeopteryx* did no more than flap heavily from tree to tree, it must have been more advanced than modern gliding animals in its range and its ability to steer to a landing place. There are birds in today's forests that manage quite well without flying far or often, and *Archaeopteryx* would not be out of place with them. There is, however, a difficulty in that there do not seem to have been any trees for *Archaeopteryx* to have flown from along the Solenhofen shore. The fossil record of the area shows it to have been open country, but there is no reason why *Archaeopteryx* or its immediate predecessors could

not have evolved in a neighbouring habitat and moved into open country where slopes and sea breezes would have assisted its flight. If this is the case, it is a good argument for *Archaeopteryx* having been a reasonably competent flier.

The next question is how *Archaeopteryx* evolved from the reptiles. There is no certain answer. It could have been related to the dinosaur *Composgnathus* with which its fossils have been confused, but there are other contenders among the large range of ancient reptiles.

Feathers evolved from reptile scales, probably to help keep the body warm. There is evidence that at least some dinosaurs were warm-blooded and would have needed an insulating coat of fur or feathers. Once the ancestral bird, often called Pro-avis for convenience, was equipped with rudimentary broad-vaned feathers, they could be adapted to provide the lifting surface that other gliders have created from webs of skin. Feathers are a great improvement; they can grow beyond the supporting skeleton and if one is damaged it can be shed and replaced.

The problem is how Pro-avis used its feathers to take to the air. There are two hotly contested theories: the arboreal, or trees-down, and the cursorial, or ground-up. In other words, Pro-avis could have evolved flight by leaping from trees and gliding down or by running on the ground and lifting off.

The cursorial Pro-avis is visualized as a bipedal dinosaur running after its prey, perhaps insects or amphibians and reptiles. It held out its feathered arms to get some lift which would make it more manoeuvrable and running would be much less strenuous, so it would have spent less time and effort in hunting. The arboreal Pro-avis is believed to have been a tree-dweller, as are many present-day reptiles. It would have launched itself into the air and glided on outstretched arms to reach the next tree. This would have been safer and quicker than climbing down one tree and up the next.

With both theories, flapping would have increased both lift and propulsion but the crucial point is how Pro-avis' successors developed the complicated wing movements and aerodynamics that make flapping flight possible. As yet, no fossils have been found to give a clue as to the identity of Pro-avis and, in the end, the choice of theory depends on opinion. Some experts believe 'trees-down' is feasible while others produce convincing arguments that 'ground-up' has theoretically a more sound basis.

One point in favour of the former theory is that modern gliding animals, with the exception of the fishes, are tree-dwellers that live above the ground. Many forest animals leap from branch to branch to escape from enemies and extend their search for food, and gliding is a useful extension of this ability. Unfortunately, the fossils of *Archaeopteryx* can be interpreted to support either theory.

Plotting the course of the evolution of birds is bedevilled by the limitations of the evidence. The fossils of *Archaeopteryx* are a sort of Rosetta stone of palaeontology that opened a previously closed subject, but the archaeologists' Rosetta stone gave vital information that led to the deciphering of many

The skeleton of Archaeopteryx *as it appeared in life, compared with that of a pigeon (not to scale). The general form is very similar but* Archaeopteryx *has the heavy jaws and teeth of a reptile, a long tail and well-developed finger bones. The pigeon has a keel on the breastbone, which anchors the large breast muscles working the wings, reinforced shoulder bones (the wishbone), rib cage and pelvis, and a reduced tail (the pygostyle).*

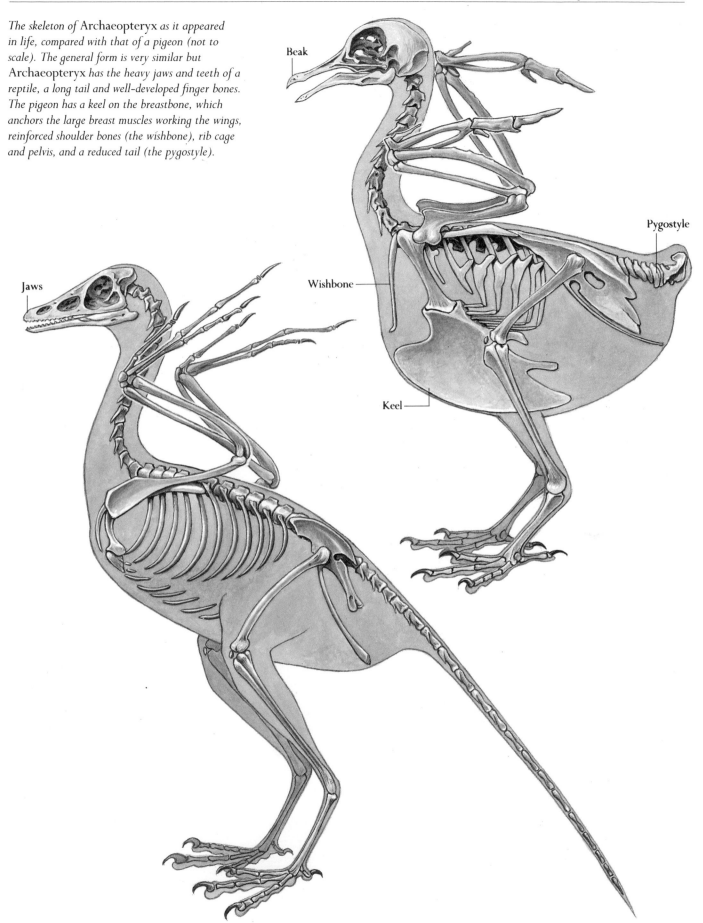

Beak

Pygostyle

Wishbone

Keel

Jaws

other inscriptions. There are no other fossils which, combined with our knowledge of *Archaeopteryx*, can help to decipher and trace the early evolution of birds, so the arguments will continue until more fossils are found.

The descendants of Pro-avis must have made many attempts to become airborne, with the failures dying out and, as yet, not being revealed as fossils. *Archaeopteryx* is the descendant of one line that was successful; other types of early bird were probably in the air at the same time as *Archaeopteryx* but, again, their remains have not yet turned up, and we do not know which one survived to become the ancestor of modern birds. It may have been *Archaeopteryx* itself, but there is another gap in the fossil record with little to show how the first birds were refined into our familiar modern birds. After *Archaeopteryx* the next fossil birds come from the Cretaceous period about 60 million years later. One of these was *Ichthyornis*, a tern-like bird, which still had teeth but is the first-known bird to have a keeled breastbone for supporting the flight muscles. By 50 million years ago many modern groups of birds had appeared, but such is the scarcity of fossils that it is still not possible to trace the relationships of many modern birds and draw up a complete 'family tree'.

The moorhen, like the coot, has a claw on its thumb when young. It is used for clambering out of the nest. In the adult bird the thumb becomes the normal feathered alula

LEARNING HOW BIRDS FLY

While the evolution of bird flight continues to puzzle modern zoologists, their predecessors had great difficulty discovering how birds fly. Many people over the centuries have watched birds and wondered how they manage to float through the air while humans fall to the ground with appalling rapidity. The idea of flying like a bird to far away places or to escape enemies with a speed and ease denied to earthbound creatures is very attractive. Such freedom leads to envy and the fantasy of imitation, which is why transformation into a bird is a common theme in myths and folk tales. There was the Greek Menippus flying to visit Zeus with the right wing of an eagle and the left wing of a vulture, and the Saxon Wayland Smith who made a suit of swan's feathers.

The first person to tackle the problem of bird flight systematically was Leonardo da Vinci. At least, he is the first to have left records of his thoughts. In 1500 he wrote: 'A bird is an instrument working according to a mathematical law,' and he set out to find the law by detailed observations. *The codex on the flight of birds* is one of a collection of Leonardo da Vinci's rough notes and jottings which contains sketches of birds and mechanical contrivances that were used as models for explaining the action of birds' wings.

In many ways Leonardo da Vinci was ahead of his time. In 1989 Colin Pennycuik wrote: 'Some ornithologists think that it is tidier to measure the wing area after joining the tips of the primaries by straight lines. However, the bird's weight is supported by the feathers, not by the gaps between them, so measure the area of feathers, and omit that of the gaps.' Nearly 500 years earlier Leonardo da Vinci had made the same observation: 'See how the interstices between the primary feathers are much wider spaces than the width of the feathers themselves. Therefore you, who would study flight, do not place in your calculation the entire size of the wing.'

The accuracy of Leonardo da Vinci's observations of birds is illustrated by the following note: 'When the bird has a small tail and large wings, it opens its wings quickly and sharply, bending in such a way that the wind, by blowing directly under its wings, raises it. And this I have observed in the flight of a young falcon above the monastery at Vaprio, to the left of the Bergamo road, on the morning of 14 April 1500.' Leonardo must have been the inventor of field notes.

Most of the notes are about the control of flight. For instance: 'The tail has movements ... sometimes it is with its ends equally lowered, and this is when the bird mounts.... But when the tail is low and the left side is lower than the right, then the bird will mount in a circular motion toward the right side.' This is quite correct – the steering action of wings and tail are observable features of bird flight – but Leonardo da Vinci was unable to resolve the problem of how birds stay airborne and propel themselves through the air. He stated: 'The bird operates in the air with its wings and tail, as the swimmer does with his arms and legs in the water.' And again:

'The wings must be rowed down and behind in order to keep the instrument up and that it may move forward.' This is incorrect.

Despite his accurate field studies and knowledge of mechanics Leonardo da Vinci had not realized the essential difference between a swimmer buoyed up by water and the bird that has to work hard against gravity to stay airborne. It was clear to his fellow countryman, the physiologist Giovanni Borelli, who wrote a century later that 'some people do blunder strangely, for they think it [flight] ought to be done as in Ships, which ... through the means of oars ... recoil at the contrary motion, and so are moved forward. In the same way they affirm that the wings are flapped with a horizontal movement towards the tail and so strike against the undisturbed air, the resistance of which occasions ... their forward motion.

'But this is repellent to the evidence of the senses and of reason, for we never see the larger Birds, such as Swans, Geese, and the like, while flying, to flap their wings toward the tail with a horizontal motion, but always to incline them downwards, describing circles set perpendicularly to the horizon. Moreover, in Ships ... there is no need to prevent their descent when they are sustained by the ... density of the water. But in the case of birds, it would be foolish to make such a horizontal motion, which would rather hinder flight as the speedy downfall of the heavy Bird would result from it.'

Borelli had grasped the essential point that birds require a force acting downwards to overcome gravity. His theory was that the downstroke of the wings could be compared with a wedge driven into a block of wood. The elasticity of the split

A sketch of the whirling arm that Sir George Cayley used for his pioneering experiments on aerofoils. The arm was rotated by a weight pulling on the cord running over the pulley.

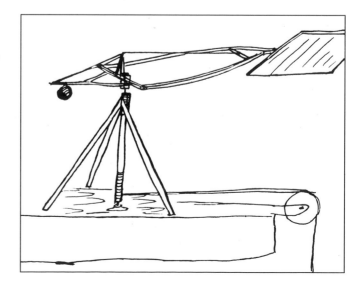

wood tends to drive the wedge out again (like an orange pip squeezed between finger and thumb); in the same way, the air pushed aside by the wings of a bird tends to push them upwards and forwards.

However, it was not until the early nineteenth century that the forces acting on a wing as it moves through the air were properly understood. Sir George Cayley, a great observer of birds, moved on from the observations of his predecessors to establish the essential principle of flight with the aid of a simple experiment. He saw that the basic problem is 'to make a surface support a given weight by the application of power to the resistance of air' and demonstrated it by sweeping a feather through the air at a slight angle and showing that it generated a lifting force. It was the first proof of the force of lift which is the basis of bird and aeroplane flight.

Calculations based on observations of the flight of herons and rooks enabled Cayley to construct, in 1853, what he called

a 'governable parachute' (i.e. a glider), fitted with paddles to provide auxiliary propulsion and extend its flight. No description of this historic machine survives, but an eye-witness account tells how Sir George's coachman was enticed to make the West's* first successful flight of a man-carrying, heavier-than-air machine. Even so, few details remain except the words of the epoch-making pilot: 'Please, Sir George, I wish to give notice. I was hired to drive, not to fly.'

The next great pioneers of flight were the Lilienthal brothers. Otto Lilienthal called his book *Bird flight as a basis of aviation* and the brothers used observations of birds to underpin the designs of gliders which they flew hundreds of times until Otto was killed in a crash in 1896. The problems encountered by the pioneering Lilienthals and the Frenchman L.P. Mouillard, another observer of bird flight, centred mainly on control and power. In 1903 the Wrights made the breakthrough into powered flight, following the invention of

The design of gliders built by the Lilienthal brothers was based on their observations of the flight of birds. Their main problem was maintaining stability in level flight.

* The Chinese Emperor Gao Yang launched gliders carrying condemned prisoners from a tower in the sixth century. They were unsuccessful, but it is reported that the Emperor gained much enjoyment.

the internal combustion engine which gave them a lightweight power unit, but uncertainties of controlling the forces of fast-moving, lightweight machines, which being heavier than air are ultimately under the control of gravity, continued to claim the lives of pioneer aviators.

Eventually, the development of aircraft progressed to the point where the aviators' knowledge of aerodynamics was applied to birds. The early issues of aeronautical journals are filled with articles by engineers and physicists using their newly acquired expertise to explain bird flight. However, the study of bird flight increasingly became divorced from the human desire to follow birds into the air. Effort went into the technology of faster and larger heavier-than-air machines and their problems bore little relation to those experienced by the flapping wings of small birds. It is often said that the Spitfire fighter of World War II was inspired by the sight of a gull soaring overhead, but there is no truth in the story; its origin lies in the evolution of a design based on the successful application of aerodynamics.

The first man to study flapping flight in detail was Etienne-Jules Marey, a contemporary of the Lilienthal brothers. He was a brilliant designer who constructed a camera that took photographs at a rate of 11 per second. His multiple-image pictures of pigeons, ducks, hawks, gulls and even pelicans are crude in comparison with modern colour photographs taken by stroposcopic flash, but they are the first to show the pattern of a bird's wingbeat. Marey also constructed pressure-recording instruments that he attached to birds to trace the movements of their wingbeats.

After Marey, most interest in bird flight was focused on relatively simple gliding flight and many years elapsed before attention came back to the more complex patterns of flapping flight. One difficulty that researchers faced in interpreting how a wingbeat propelled and supported the bird is that the movement of the wing varies constantly, even between two successive beats. Eventually, in the 1950s, R.H.J. Brown of Cambridge University used electronic flash to 'freeze' the wingbeats of a pigeon at 100 frames per second. He was thus able to demonstrate clearly the exact sequence of events in a wingbeat and to deduce, partly from the bending of the feathers, the generation of forces that propel the bird.

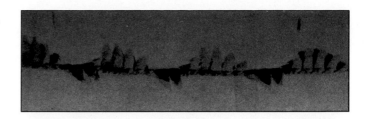

Over a century ago, the Frenchman Etienne-Jules Marey was taking multi-image photographs of birds in flight to study the pattern of their wingbeats. Repeated exposures at very short intervals reveal successive positions of the wings.

Much about bird flight remains unexplained because conventional aerodynamics deals with the fixed wings of aeroplanes and does not relate to the complex wingbeats of birds' flapping flight. The mechanics of bird flight is now the province of biologists, but the uncertainties in these pages will show how difficult the study has proved to be. Birds have been trained to fly in the laboratory, where instruments can be used to record their performance in level flight. But follow a party of rooks as they fly past, ducking and weaving apparently for no purpose other than enjoyment, thrill to the sight of a sparrowhawk snatching its prey out of the air or watch a close-packed flock of starlings wheeling over its roost: it is impossible to explain the forces acting on their bodies and the way these are controlled. We are still ignorant about much of the physics of bird flight but there are enough parallels with aeroplanes to help explain some of the aerodynamic tricks used by birds in their mastery of aerial life.

CHAPTER 2

THE BIRD AS A FLYING MACHINE

A basic knowledge of aerodynamics and the way a wing creates lift is necessary for understanding the facts of bird flight. In this chapter the simple aeroplane wing is used as a model, first to explain how birds glide and then to explore the complexities of flapping flight.

Flying is the most strenuous activity in the animal kingdom, so birds combine high power output with low weight. Their bodies are adapted to meet these requirements. The skeleton and internal organs have been lightened, but the flight muscles are huge and a unique breathing system provides them with oxygen.

Cape gannets display effortless grace as they patrol over the sea in search of fish off the coast of South Africa.

BASIC PRINCIPLES

The slow, deep wingbeat of a heron or even a crow gives the impression that the bird is 'rowing' through the air. It is not surprising that Leonardo da Vinci and many others thought that birds' wings were super-paddles mysteriously pushing a mass of air back and down so that the bird is buoyed up and propelled forwards. It is an unfortunate impression that lured several pioneers to an early death in totally useless 'ornithopters' – man-carrying machines that were supposed to imitate birds – and it delayed the true understanding of how birds fly. A bird does not flap its wings to push air down and its body up. It moves them forwards through the air so that a flow passes over their surface and generates a lifting effect. An aeroplane's wings work on the same principle.

The wing of an aeroplane is subject to four forces as it travels through the air and, if it is moving on a straight and level course, each pair of forces must be equal and opposite: weight acting vertically downwards under the pull of gravity is balanced by a lifting force generated by the wing, and the forward thrust (provided by the aircraft's engines) is held back by the resisting force of drag.

An aeroplane flies because its wings generate enough lift by its forward movement through the air to overcome its weight. The aerodynamic principles can be demonstrated quite simply. A wing is not a flat board; it is convex above and concave below, both curves tapering gently to the trailing (rear) edge. This shape is called an aerofoil and its curvature is the camber.

As the airstream meets the leading (front) edge of the wing, it parts to pass over and under the aerofoil. The air passing over the upper surface is forced to speed up over the camber and its pressure drops. The reason for the pressure drop was first explained by the Swiss physicist Daniel Bernoulli (1700-82). He found that when water flowing down a pipe meets a constriction it speeds up (think of a river flowing through a gorge) and its pressure drops. This sort of pipe is called a venturi tube. The constriction in the pipe is shaped rather like the camber of a wing and exactly the same speeding up of flow and decrease in pressure takes place in the airflow over the wing.

You can test the Bernoulli effect and the lifting force of an airstream over a wing with a sheet of typing paper. Hold it by two corners so that it hangs down at the back and blow hard so that your breath hits the top of the front edge. You will see the paper rise in the airflow because of the reduction in air pressure on its upper surface.

The reverse effect takes place on the underside of an aerofoil: the air slows down and pressure increases. You can think of the lift coming from the reduced pressure over the upper surface of the aerofoil 'sucking it upwards' and the increased pressure underneath 'pushing it upwards.' The amount of lift depends on the degree of curvature because a steeper curve diverts the air farther and makes it travel faster, as Sir George Cayley realized when he examined a heron and wrote: 'I am apt to think that the more concave the wing to a certain extent the more it gives support.'

Another way of considering the lifting force is not as low pressure 'pulling' the aerofoil up, but as the flow of air being directed downwards behind the wing. According to Newton's Third Law – 'For every action there is an equal and opposite reaction' – the movement of the mass of air downwards produces an equal and opposite force keeping the wing up.

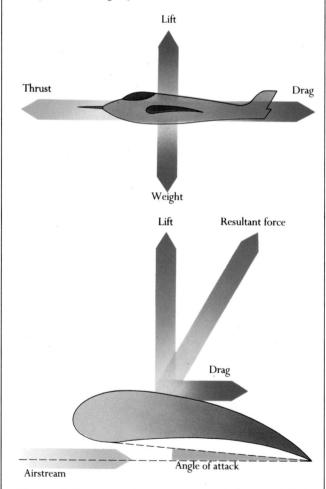

THE FOUR FORCES

When an aeroplane, or bird, is in level flight, four forces are in balance. Lift and weight, and thrust and drag, are equal and opposite. Changes in lift or thrust cause the aeroplane to change speed and altitude.

The angle between the airstream and the aerofoil is called the angle of attack. The flow of air over the aerofoil generates lift and drag which combine to create a resultant force. As shown here, lift is greater than drag and the resultant is more vertical than horizontal.

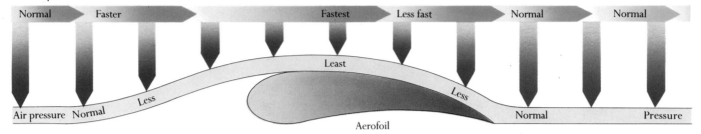

Airflow speed

Normal Faster   Fastest   Less fast   Normal   Normal

Least

Less   Less

Air pressure Normal   Normal   Pressure

Aerofoil

ABOVE The Bernoulli effect. Fluid flowing down a pipe accelerates when it passes through a constriction and its pressure drops. This is what happens when air flows over an aerofoil. The fall in pressure acts as a lifting force.

BELOW The lifting force of a current of air flowing over an aerofoil can be demonstrated by blowing hard across a sheet of paper and watching it rise.

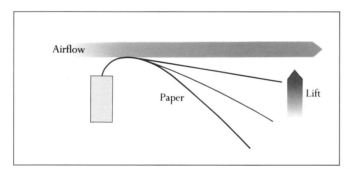

Airflow

Paper   Lift

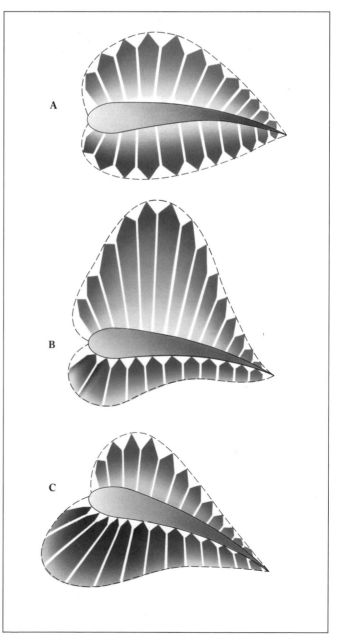

A

B

C

RIGHT A schematic representation of the forces acting on an aerofoil. As air flows over the aerofoil it creates areas of high and low pressure which exert forces whose direction and strength is indicated by the direction and size of the arrows. At 0° angle of attack (A), low pressure on both sides of the wing almost cancel out. The optimum angle of attack is about 5° (B), where pressure is very low on the upper surface and high under it generating lift. At about 15° the aerofoil stalls (C) - the reason for this is that the low pressure on the upper surface decreases while pressure is building up underneath causing high drag (see overleaf).

The wing is subject to a second force, the drag which tends to pull it back in the same direction as the airflow. Drag is the force which you experience if you put your hand out of a window of a moving car. Part of the drag on a wing is a direct result of the production of lift as the wing travels through the air. This form of drag is called induced drag. High pressure air under the wing tries to flow around to the low pressure area on the top, forming a swirling vortex which spirals along the trailing edge to the wingtip where it is shed as the wingtip vortex. Wingtip vortices can sometimes be seen as white trails

The airflow over an aerofoil generates forces of lift and drag. The resultant of these two forces – the lift/drag ratio – depends on the angle of attack. At 0°, lift and drag are low; as the angle of attack rises, lift initially increases faster than drag so the lift/drag ratio is a maximum at an angle of attack of about 5°. Thereafter, drag increases rapidly and, at an angle of attack of about 15°, the aerofoil stalls because the airflow over the upper surface becomes turbulent, so that lift disappears but drag builds up.

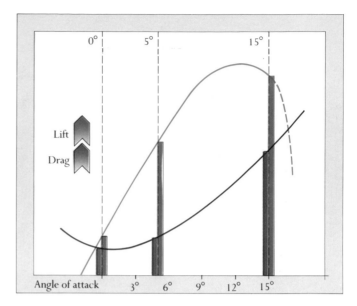

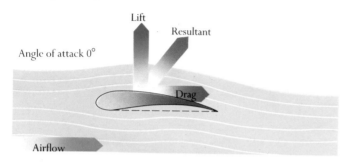

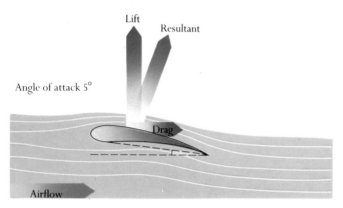

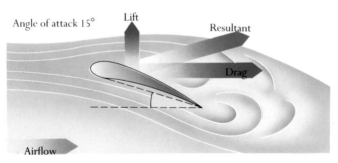

on airliners as they take off and the whirling effect appears momentarily on the spoilers of racing cars. In both instances they are most visible in damp conditions. Reduction of wingtip vortices and induced drag is important in the design of aircraft and bird wings, as will be discussed later. Induced drag is greatest at low speed but there are two other forms of drag which increase with speed. Parasite drag is caused by airflow over the body, so streamlining becomes important at high speed. Profile drag is caused by friction of the airflow on the surface of the wings; how much it affects flying performance depends on the shape of the wings.

Lift and drag combine to give a resultant force called the lift/drag ratio, which describes the net effect on the wing. If there is 10 times as much lift as drag, the lift/drag ratio is 10:1.

An aeroplane or bird needs as high a lift/drag ratio as possible. One way of improving it is to reduce drag by streamlining the shape of the wings and body, but it is more important to control the angle that the wing makes with the airflow – the angle of attack. When the wing is horizontal to the airflow there is little difference between lift and drag, but as the angle of attack increases, the speed in the airflow over the upper surface rises and lift increases. However, beyond a critical angle, about 12 to 16 degrees, lift slumps because the airstream no longer flows smoothly over the wing but becomes turbulent and the pressure reduction is lost. At the same time drag increases immensely. The wing is said to stall and if this happens to an aircraft it falls out of the sky unless immediate steps are taken to create a normal airflow and restore lift. The aim, therefore, is to fly with an angle of attack that gives the best lift/drag ratio. As the diagram shows, this is about 3 to 4 degrees.

The wing has two other basic features which affect its flight characteristics: wing-loading and aspect ratio. The wing-loading is the ratio of an aeroplane's weight to its wing area. An aeroplane with a high wing-loading has to fly faster to generate the lift to stay airborne, so a heavy, small winged fighter plane has a high wing-loading, while a sailplane has a low wing-loading. The aspect ratio is the ratio of wing length (span) to width (chord). A long, narrow wing has a high aspect ratio; it also has low induced drag because there is a large area for generating lift but a slender tip where the vortex is formed. In a sailplane the vortex is further reduced by making the tip pointed, which cuts down the pressure differences between the upper and lower surfaces. The low drag makes a high-aspect-ratio wing efficient and is a feature of sailplanes, but a long wing is structurally weak so aircraft designed for violent aerobatics have short wings and low aspect ratio. This is an important consideration in the adaptation of birds for different ways of life.

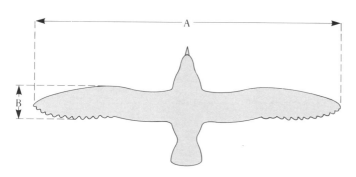

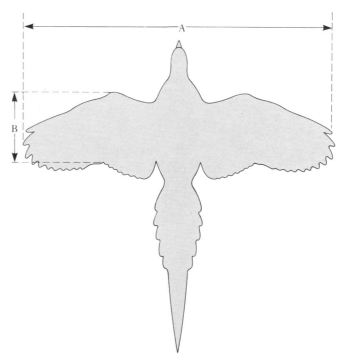

Aspect ratio is the ratio between the wingspan (A) and the chord or breadth of the wing (B). A fulmar with long, narrow wings (ABOVE) has a higher aspect ratio than a pheasant with short, broad wings (RIGHT). High-aspect-ratio wings are aerodynamically more efficient, but they cannot be flapped as rapidly as short, broad wings of the same wing area.

TOP OF PAGE Its long, slender wings, with pointed tips, give the great frigatebird a very efficient, buoyant flight that enables it to fly for long periods over the sea.

THE WING OF A BIRD

Compared with the simple aerofoil of an aeroplane, the wing of a bird is a complicated piece of design. The essential difference is that it functions both as a lifting surface and a propeller, so it must be rigid enough to withstand considerable forces of lift and drag yet remain flexible enough for the complexities of flapping flight. And when not in use, it folds out of the way.

The wing evolved from the basic reptile forelimb with five toes used for walking. Once freed from this ancestral function, the birds' forelimb has undergone radical changes in the structure of its bones. The shoulder has moved up so that the bird's body, and its centre of gravity, hangs below the wings to confer stability. The humerus (upper arm bone) is usually short, particularly in birds with a rapid wingbeat, such as hummingbirds, but is long and slender in gliding birds (if an albatross flapped its wings like a hummingbird they would snap). The shoulder is a 'universal' joint, which means that the humerus can rotate in the complex movements of flapping flight and folding the wing. In contrast to reptiles and mammals, however, its main movement is up and down for

flapping, rather than to and fro for walking. The elbow and wrist joints (the latter known as the carpal joint) are more rigid and move, like hinges, mainly in the horizontal plane for spreading and folding the wing.

The greatest anatomical change in the bird's wing has taken place in the wrist and hand. Most of the bones have fused together or have disappeared entirely. The main survivor is the carpometacarpus which represents the bones in the palms of our hands. Only three digits remain and they, too, have lost most of their bones. The thumb and third finger are each represented by a single bone and the second finger has only two bones.

The flight feathers, or the remiges, are attached to the wing bones. The primary flight feathers, or pinions, usually 11 in number, are attached to the second and third fingers and the wrist bones. They provide the greater part of the wing area and are essential for flight. The secondary feathers are attached to the ulna, one of the bones in the forearm. Their number depends on the shape of the wing and ranges from six or seven in hummingbirds to 40 in albatrosses. If a pigeon's secondaries are clipped, reducing its wing area by half, it can still fly, but removal of no more than the tips of the primaries grounds it. There are also a few tertials on the humerus of the upper arm, as well as three or four feathers on the thumb, called the alula or bastard wing, which are used in the control of flight at low speed (see page 33). The remaining feathers are called coverts. They overlap like tiles on a roof to make the curved, streamlined shape of the aerofoil.

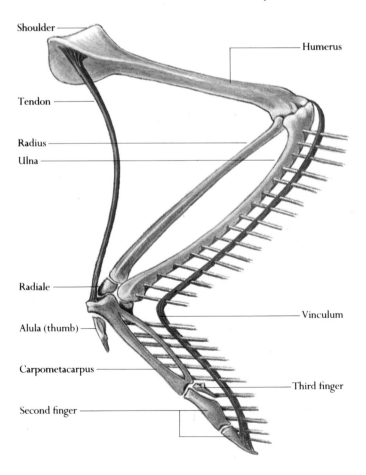

Shoulder
Humerus
Tendon
Radius
Ulna
Radiale
Vinculum
Alula (thumb)
Carpometacarpus
Third finger
Second finger

ABOVE A gannet hangs in the air at its cliff-face colony, showing how the inner and outer wings articulate at the wrist.

LEFT The main bones and tendons of a bird's wing. The tendon running from the shoulder to the wrist stops the wing opening too far and stretches the web of skin forming the leading edge of the wing. A ligament, the vinculum, holds the flight feathers in place.

The arrangement of flight feathers on the wing. The secondaries of the inner wing are attached to the ulna and the primaries of the outer wing are attached to the bones of the hand. The alula feathers are borne on the thumb.

Secondaries

Alula

Primaries

ABOVE Section of the wing showing how the contour and flight feathers on the bird wing make a smooth aerofoil outline, similar in shape to an aeroplane wing.

SPREADING THE WING

The mechanism for spreading and folding the wings is very neat. When a bird opens its wings the bones are straightened and the primary feathers are automatically spread into the flying configuration. You can try this with a dead bird. Put your thumb behind the elbow and gently push. As the elbow joint extends, the wrist stretches out with it and the primaries are fanned. The movement is due to the wing bones acting as a series of levers, like the draughtsman's parallelogram. As the humerus swings out it pulls the radius bone of the forearm inwards. The outer end of the radius is connected to the carpometacarpus and the tiny radiale bone, so these are pulled over the rounded head of the ulna to straighten the wrist joint.

Eventually the wing will spread no further, although it is not completely straight (feel the bones: they are still bent in a shallow N). Further extension has been stopped by a tendon running from a muscle on the shoulder to the wrist. The same tendon stretches the web of skin (the patagium) on the leading edge of the wing and shapes the camber on the anterior part of the aerofoil.

The bases of the primaries and secondaries are joined by an elastic ligament (the vinculum), which runs between the elbow and the tip of the second finger. When this is stretched by the spreading of the wing, the flight feathers are automatically set in place, roughly at right angles to the bones to which they are attached. While the wing is spread the primaries can be spread further, or partly folded back, by muscular action, but some birds have a flange on the carpometacarpus which slides over a projection on the ulna to lock the wrist in place.

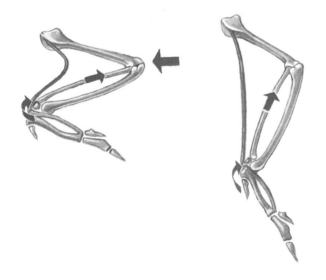

ABOVE The mechanism of wing-spreading. As the humerus swings out and the elbow joint opens, the wrist is automatically extended and the flight feathers spread.

FEATHERS

Feathers are a marvel of construction. They combine lightness, strength and flexibility. If you take a pigeon's flight feather and bend it double, it will flick back into shape when released. Feathers, like mammals' fur, are made of keratin, and they perform the same functions of keeping the animal warm and dry and protecting it against injury. They probably evolved for these purposes, but they now also perform the unique functions of streamlining the body and providing lifting and control surfaces on the wings and tail.

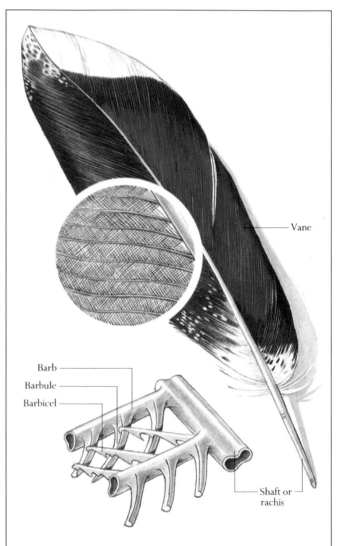

FEATHER STRUCTURE

The vane of a flight feather is made up of rows of barbs on each side of the central shaft or rachis, each one sprouting hundreds of barbules which overlap and are hooked together by millions of microscopic barbicels. The strength and flexibility of the feather are due to the barbicels forming firm but mobile links.

Feathers have the advantage over the skin-membrane wings of bats and pterosaurs because they give the wing a modular construction which limits damage. A tear in a wing membrane is serious, but a split feather can be repaired by preening and lost feathers are quickly replaced.

The wing and tail feathers that are important in flight are sometimes called vaned feathers. They consist of a central shaft or rachis with webs on each side that make up the flat vane. The vane is composed of hundreds of pairs of branches called barbs. Each barb is joined to its neighbours by several hundred barbules which overlap in a herringbone pattern and cling to each other with tiny hooks or barbicels. The millions of hooks give the vane its strength and also confer its flexibility because they let the barbules slide over each other.

The fine structure of the vane makes it waterproof and almost windproof. It also means that a single layer of overlapping feathers makes a very light but effective wing surface. It might seem that the surface of a feathered wing is not as streamlined as the smooth skin of a bat's wing membrane, but experiments show that the rough, regular pattern of the vanes improves lift.

If a bird flies close by, you can often hear the noise of its wingbeats. A flock of sparrows makes quite a loud whispering and a flock of starlings is deafening. The constant sound must make flying like riding a motorcycle and poses the question whether the continuous noise blankets sounds important to the birds. Presumably it does not to any significant level, or more birds would have developed the muffled wing feathers found on owl wings. The barbs on owl primaries are unhooked at the tips, giving the feathers soft, frayed edges, so a gliding owl is not only much quieter but the sound of its passage through the air is of a low frequency that will go unheard by small mammals and birds. The drawback is that the wings have more drag and flying is more of an effort. Day-flying owls which do not rely so much on stealth have normal feathers.

ABOVE A pigeon's feather shows the parallel barbs which are linked by the barbules in an overlapping herringbone pattern to make an almost windproof fabric.

BELOW Unhooked barbs make loose fringes on the leading edges of an owl's flight feathers to deaden the noise of its wingbeats.

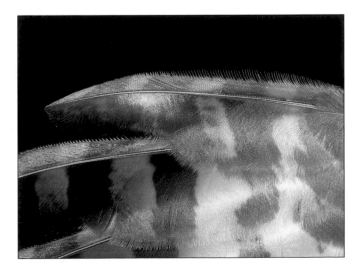

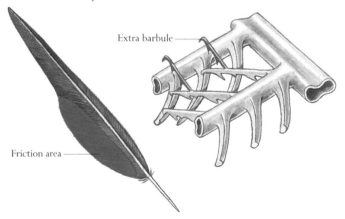

Extra barbule

Friction area

Flight feathers have friction areas where barbules are supplied with extra barbicels which hook onto neighbouring feathers. When the feathers are spread, the friction areas hold them in position.

Shaft   Vane

GLIDING

A gliding bird is sufficiently like an aeroplane to be described in the same aerodynamic terms. The bird uses the pull of gravity as 'thrust' (as a cyclist freewheels downhill) to overcome the backward drag and propel itself forward, slicing the wings through the air with sufficient speed to generate enough lift to hold it up. If there was no drag, the bird would glide without losing height and stay up indefinitely, but in practice, drag slows it down so it sinks through the air. To extend the glide as long as possible, the bird will want to minimize the glide angle (the angle between horizontal and the flight path) and consequently the sinking speed (the rate of losing height). It sets its wings at the angle of attack that will give the best lift/drag ratio. A low forward speed helps, but the bird must glide fast enough to avoid stalling. The pilot of an aeroplane that has run out of fuel and is trying to reach a runway has the ticklish task of staying airborne as long as possible by extending his glide. However, he does not fly at the minimum gliding speed because the plane would become unstable and liable to stall; instead, he puts the plane's nose down and glides a little faster, at the best gliding speed where drag is at a minimum, thereby increasing the distance he covers.

A bird has the edge on the hapless pilot because it can change the shape of its wings to alter their aerodynamic characteristics. The 'variable geometry' of a bird's wings gives it much greater control over gliding flight by maintaining the best lift/drag ratio and minimum gliding angle through a range of speeds. The pilot attempting to glide can do little more than raise or lower the nose of his aeroplane, altering the gliding angle to vary his speed, but a bird spreads its wings to full stretch to increase their area, thereby reducing the wing-loading and giving a slower gliding speed. Closing the wings

BELOW A bald eagle dives steeply at speed with its wings half closed to reduce the wing area and lose lift.

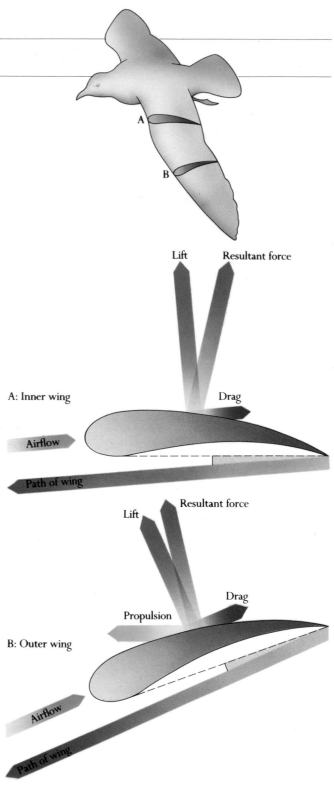

ABOVE The flexible construction of a bird's wing gives it advantages beyond the limitations of the rigid aeroplane wing. The angle of attack can be varied along its length to obtain the optimum lift and thrust. In this example, the more rigid inner wing provides lift mainly, but the outer wing has twisted in the airstream on the downstroke and is angled downwards so that the force it exerts is directed forwards to pull the bird through the air.

To slow down and maintain control while gliding, the sooty albatross spreads its tail and lowers its feet as airbrakes.

has the opposite effect. A pigeon glides slowly at 30 kilometres (19 miles) per hour with its wings fully spread to their maximum area. To speed up, it progressively folds its wings at the wrist so that the primaries fold and the wing area is reduced until, with the area almost halved, gliding speed is increased to 80 kilometres (50 miles) per hour. I see this most spectacularly on winter afternoons when a large flock of wood pigeons returns to its roost in a nearby wood. The pigeons fly over at a considerable height, then begin a rapid, kaleidoscopic descent as individuals drop with folded wings, then check, glide in an arc with wings fully spread, and drop again until they are level with the tops of the trees.

Long-winged birds glide well by adopting a 'nose up' position with a very small gliding angle and avoid stalling by twisting their wings to reduce the angle of attack. Moreover, the angle of attack can be varied along the length of each wing. A slow-gliding bird can sometimes be seen to have the inner wing, bearing the secondaries, at a high angle of attack but the

hand, bearing the primaries, twisted so that it is almost flat or even turned down. The elasticity of the feathers also enables the wing to change shape automatically to compensate for variations in the airflow. The result is that birds can glide at speeds which would be too slow to support them in the air if they had the inflexible wings of an aeroplane.

Added to the variable geometry of the main structure of the wing, birds have several devices for further increasing its aerodynamic efficiency. The rudimentary thumb or alula is not a useless vestigial organ but has been retained to prevent stalling at low speed. A stall starts when the angle of attack rises and the smooth flow of air over the wing breaks up and becomes turbulent. The alula bleeds a current of air across the top of the wing, maintaining the smooth flow and preventing the start of turbulence. The angle of attack can then be raised to give greater lift at low speeds, and the bird can fly more slowly without the danger of stalling. The neatest feature of the alula is that its action is automatic: a high angle of attack creates low pressure on the leading edge of the wing and the alula is sucked out.

The alula is a good example of the 'Nature thought of it

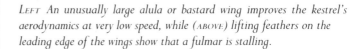

LEFT An unusually large alula or bastard wing improves the kestrel's aerodynamics at very low speed, while (ABOVE) lifting feathers on the leading edge of the wings show that a fulmar is stalling.

THE ALULA

The slat on an aeroplane wing and its equivalent, the alula, on a bird wing reduces the chance of stalling. As the angle of attack increases, the airflow over the wing becomes turbulent and lift is lost. The slat bleeds air through a slot to re-establish a smooth airflow.

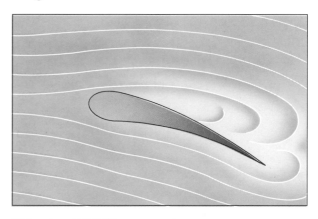

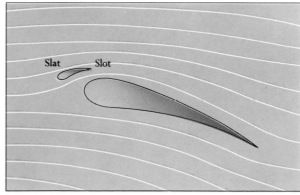

Slat   Slot

first' story, because birds were using it to prevent stalling at low speed for millions of years before Sir Frederick Handley-Page invented exactly the same mechanism to prevent aeroplanes from stalling. In this instance the aircraft designer had not watched birds for inspiration but admitted that the invention would have come much sooner if he had realized what birds were doing. The device on an aeroplane's wings is called a slat (with the gap between it and the wing called a slot) and is used in the same circumstances as the alula: on landing so that the aircraft can touch down at low speed without going out of control or risking damage to the landing gear that might come with a fast landing.

Wingtip slots are another device that can be brought into action when flying slowly. The vanes of the primary wing feathers of many birds are reduced in width in a distinct step near the tip. This is known by ornithologists as emargination or notching. When the wing is spread, the tip separates into 'fingers' with gaps or slots between the primaries. There have been several theories about their exact function but it is still not clear how they work. There may, indeed, be more than one function. It is, however, indisputable that wings with pronounced slots are found on birds with relatively short, rounded wings, whereas those with long pointed wings, such as falcons, waders and ducks, may have only one small slot.

It was mentioned earlier that a long, pointed wing produces less induced drag, but for a variety of reasons this design is not always possible. Wingtip slots are a feature of large birds such as vultures, cranes and pelicans which have broad wings of low aspect ratio. These are typically species which fly or glide at low speed yet cannot afford long wings as they would make take-off difficult. Another group, including the pheasants for example, are woodland birds which would be hampered by long wings yet need high power at low speed for taking off.

Above: A bald eagle glides slowly by spreading its wings to their full extent. The primary flight feathers separate into 'fingers' which are twisted and staggered by the airflow to make the wings more efficient at low speed.

Below Separated primaries may act as individual winglets to counteract stalling. A high angle of attack on the main wing, for instance when a bird is landing, causes it to stall, but the primaries twist in the airflow, reducing their angle of attack and continuing to provide lift and thrust.

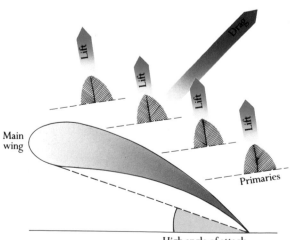

Main wing

Lift

Drag

Primaries

High angle of attack

One suggestion is that each emarginated primary acts as a small, independent aerofoil or winglet which twists in the airstream. Because the rear part of the vane on an emarginated feather is larger than that in front of the shaft, the feather is twisted to reduce its angle of attack. So, when the angle of attack of the main wing rises to stalling point, the winglets keep a better angle of attack and maintain lift. It has also been suggested that slotting reduces induced drag; this would lower the rate of sinking in a gliding bird and delay the need for flapping to maintain height. This would seem to be the case in the display flight of the collared dove, which ends with a wide, gliding circle. The stretched wings are eye-catching, which is the object of the exercise, and the effect is prolonged because the slots enable the dove to float in the air. At the end of the display it relaxes its wings, reducing their span, the slots close and the dove drops rapidly back to its perch.

THE TAIL

Birds that lose their tails still manage to fly quite well, yet the tail has an important function in controlling flight, steering and braking. The spread tail also acts as an extra lifting surface and reduces the overall wing-loading, thereby assisting slow flight. The wing-loading of birds of prey is reduced by around 20 per cent when the tail is spread. However, possibly the main function of the tail for most birds is to improve the flow of air over the wings at low speed. The spread tail forms a slot behind the wing which sucks air over the inner section of the wings so that the flow does not become turbulent at high angles of attack, such as at take-off and landing, when the tails of birds are most conspicuously spread. Frigatebirds, terns and swallows, which fly slowly but have high-aspect-ratio wings without wingtip slots, have forked tails which they spread into long flaps to aid low-speed lift generation. The puffin and other auks and the divers, which have high wing-loadings and find it difficult to fly slowly when coming down to land, get extra lift by lowering and spreading their webbed feet to supplement their particularly short tails.

A female snowy owl takes off with her widely spread tail overlapping her wings. Acting like the flaps of an aeroplane, the tail increases the area for lift and improves the airflow over the wings.

FLAPPING FLIGHT

Sustained flight requires a thrust force, an expenditure of energy sufficient to overcome drag and keep the bird on a straight and level course. All true flying animals have achieved this by flapping their wings for both lift and propulsion. Understanding this point was Sir George Cayley's major contribution to the development of the aeroplane, because he realized that the two functions of lift and propulsion would have to be separated for flying machines to be successful.

With the advent of high-speed photography and slow-motion filming the elaborate sequence of movements in flapping flight was revealed and it became possible to analyse the functions of different parts of the wing at each stage of the wingbeat cycle. The following description of flapping flight is only an approximation of the precise form and function of wingbeat cycles whose details are not only extremely complex but vary between species.

A gull flying steadily at about 40 kilometres (25 miles) per hour provides a good example of the wingbeat cycle in fast flight. At the start of the downstroke the wings are fully extended at an angle of about 30 degrees above horizontal. They beat down vertically and the forward movement of the

bird sweeping the wings through the air generates lift along the entire length of the wings. At the end of the downstroke, which is not far below horizontal in the gull, the wings fold slightly and the primaries close like a fan. There is no propulsive effort on the upstroke, as shown by the absence of curvature on the primaries, but on larger birds the inner wing continues to act as an aerofoil to obtain some lift. At the end of the upstroke, the primaries generate a little lift which helps to raise and extend the wing.

One difficulty in visualizing the interaction of wing and airflow is that pictures show the wing's movements in relation to the bird's body but the important point is their motion in relation to the airflow. When the bird flaps its wings up and down relative to its body they are effectively slicing through the air and meeting the airflow head-on. This important point was first demonstrated by Etienne-Jules Marey in the 1860s. He photographed crows with strips of white paper glued to their wingtips showing that the wings are swept forward and

A flock of laughing gulls flies along the shore on easy wingbeats, demonstrating all the stages of the wingbeat cycle.

Downstroke	Upstroke

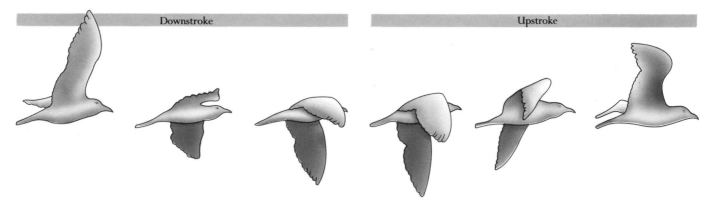

ABOVE The wingbeat cycle of a gull in steady flight. The main lift and propulsion come on the downstroke. The wings fold slightly on the upstroke and bring the wing back to the starting position for the next downstroke with the generation of a small amount of lift.

BELOW A multi-image photograph showing three stages of a coal tit's wingbeats. First, the wings are partly folded on the upstroke. Then the wings are just about to fold at the start of the next upstroke. Finally, the wings are swept forward on the downstroke.

downward, thereby proving that the birds are pulling rather than pushing themselves through the air.

A further complexity is that a bird's wing is made up of two parts which have different functions in the wingbeat cycle. The inner wing, from shoulder to wrist, carries the secondary and tertiary flight feathers and is relatively rigid, while the outer part comprising the wrist and fingers bearing the primary flight feathers undergoes considerable changes in shape, as many of the photographs on these pages show. The inner wing is beating almost straight up and down and the airflow over it comes from the movement of the bird through

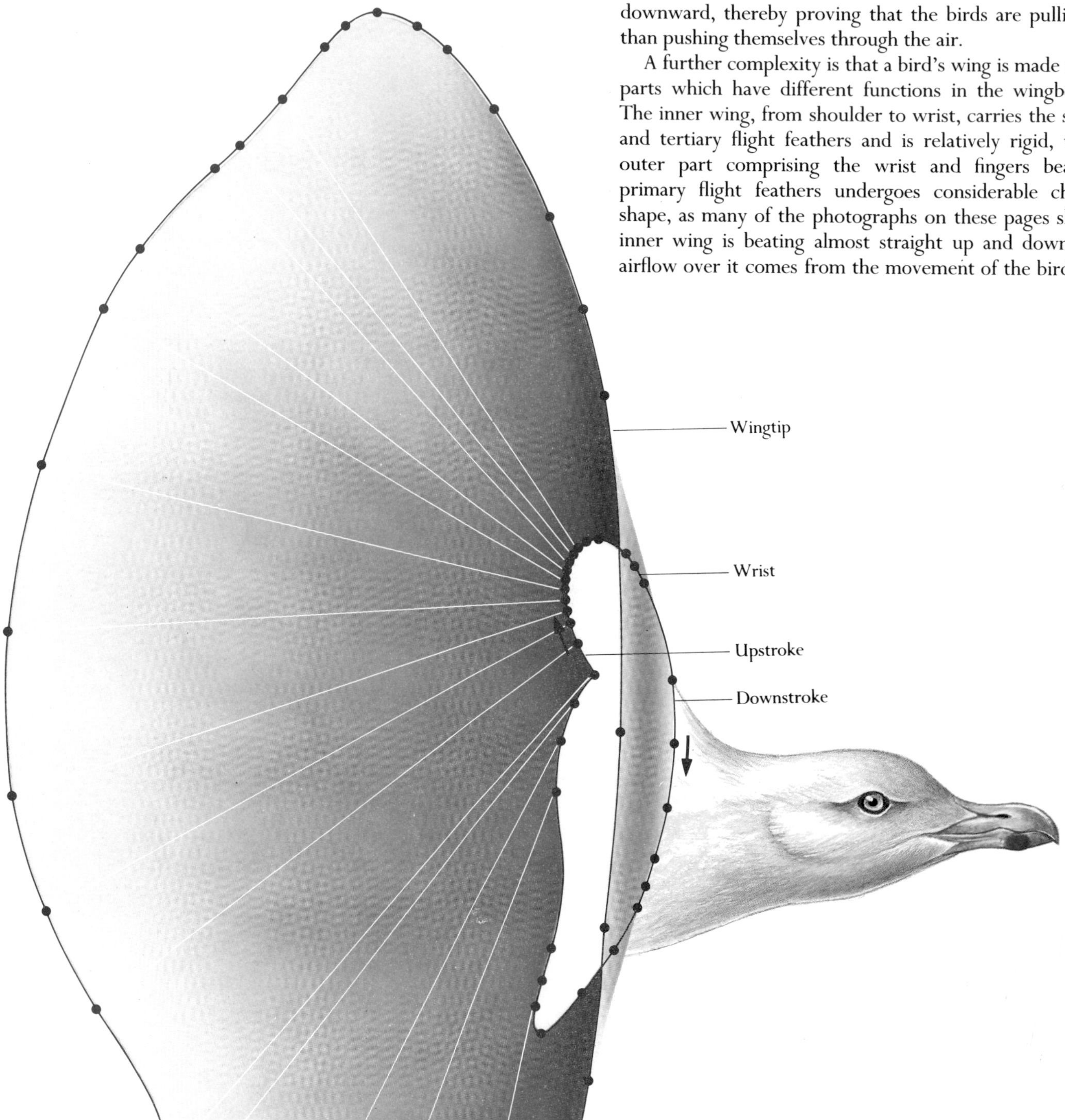

Wingtip

Wrist

Upstroke

Downstroke

The paths described by the wingtip and wrist during one wingbeat cycle show how the wing works in two sections. The dots are time-points. They show that the wingtip moves faster on the upstroke (the dots are farther apart). Lines joining the dots show the movement of the wrist relative to the wingtip. On the upstroke the inner wing is raised before the outer wing, which flips up at the end of the stroke.

the air. The resultant force is near vertical so it is producing lift with little thrust. The outer wing produces a resultant force which is directed forward to provide most of the thrust.

WINGBEATS PER SECOND

Wingbeat is not only variable, depending on the bird's speed and load, but difficult to record. These figures give an indication of the range of frequencies.

Grey heron	2
Herring gull	2.8
Starling	5.1
Pheasant	9
Mockingbird	14
Tits	25–27
Ruby-throated hummingbird	80

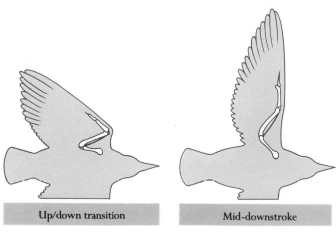

ABOVE Upstroke – the inner wings are raised and the primaries of a scops owl are about to be flicked straight at the end of the upstroke.

LEFT Downstroke – the primaries of a green woodpecker's wings bend as they are forced down at the start of the downstroke.

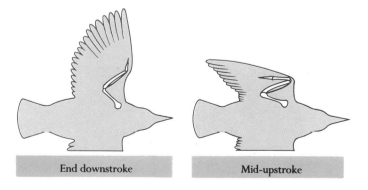

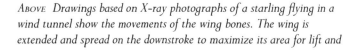

Up/down transition	Mid-downstroke	End downstroke	Mid-upstroke

ABOVE Drawings based on X-ray photographs of a starling flying in a wind tunnel show the movements of the wing bones. The wing is extended and spread on the downstroke to maximize its area for lift and propulsion. On the upstroke it is partly folded and the flight feathers close like a fan to reduce resistance as the wing is raised for the next downstroke.

POWER FOR FLIGHT

Zoologists calculate the energy budgets of animals with incomes and expenditures measured and balanced in joules.* An animal's energy income comes from the food it eats, with a small bonus from the sun, and it is spent on daily activities, such as travel, finding food and rearing a family, and in maintaining the vital functions of heartbeat, breathing and digestion. The survival of all animals depends on maintaining a positive energy budget. The more a bird can keep its 'bank account' of energy reserves in credit, the more energy it will have to spend on growth and reproduction.

Flight uses so much energy that the cost of flying is crucial to a bird's energy budgeting and a very important factor in its style of living. Although flight gives birds immense advantages, there are drawbacks and about 40 species have given up flying altogether (as well as the great auk, dodo and others which have become extinct in the recent past); many more, the gamebirds and rails for instance, fly only rarely. Mankind may envy birds their freedom, but we should remember that flight can be a handicap.

When considering the role of flight in the life of a bird, it is useful to think of the implications of the energy requirements. First, birds do not fly if they do not have to. If a starling at the bottom of the garden sees a crust thrown from the kitchen door, it is faced with the choice of walking or flying to get it. Walking is less of an effort but, if there are other starlings nearby, there will be a race for the food. Speed is of the essence so the starling flies up the garden.

The energy balance can only be ignored for a good reason. Birds go into energy debt briefly through the exertion of chasing a meal and they throw caution to the winds to escape from a predator. In the long term, the imperative to rear offspring may lead some birds to lose weight during the time that they are supplying their growing nestlings with food.

Flapping flight is one of the most strenuous activities in the animal kingdom and requires a continuous delivery of power to the wings, perhaps for hours on end when migrating, that would leave an earthbound animal exhausted. The amount of power that a bird's flight muscles can generate is proportional to their weight, and to double the power output of a muscle its weight must be quadrupled. This means that heavier birds find flight more difficult than light birds and that there is an upper limit to the weight of a flying bird, which seems to be about 10 to 15 kilograms (22 to 33 pounds). The birds that come nearest to this are the mute swan, white pelican, Californian condor and kori bustard. They have little surplus power over that required for steady flight and consequently have difficulty taking off. They also lack manoeuvrability, which explains why swans so often collide with power lines. It would be simple to drop under a cable, but the swans try desperately to gain height to clear it.

---

*Dieters are more familiar with the calorie as a unit of energy. One calorie = 4.13 joules.

BELOW The kori bustard, one of the heaviest flying birds, takes off and flies with reluctance.

POWER AND SPEED

The energy used or, in other words, the fuel consumption, is so crucial to understanding the role of flight in the lives of birds that it is worth looking at the basic facts of power expenditure through the relationship of power and speed in bird flight.

The graph demonstrates that the relationship between power and speed is, at a simple level, little more than common sense. When a bird hovers or flies very slowly, it requires a high induced power to remain airborne by flapping because there is no forward speed to generate lift through air flowing over the wings. As its forward speed increases, lift is generated by air flowing over the wings and less energy is needed to

Like many birds that swim under water, the king cormorant of South America has small wings but they deliver high power at low speeds.

support the bird's weight in the air, so less induced power is required from the wingbeats. Parasite power is zero when the bird is stationary (i.e. when it is hovering), but is required when forward speed increases and drag on the body builds up. The profile power needed to overcome the drag on the flapping wings is extremely difficult to calculate, and although it rises slowly with increasing speed, it is usually assumed that it remains constant. The total power needed to fly at any speed is the sum of the three power requirements, as can be seen by examination of the power-speed curves.

The precise shape of the power-speed curve for each bird species depends on the individual characteristics of weight, wing shape and wingbeat pattern and muscle mass. Our interest is that it can be used to explain the relationship between the physical attributes of the bird and its flight style. Flying slowly requires mostly induced power, so birds that fly slowly need wings that produce lift but require little induced power to overcome drag. This is fulfilled by the long, pointed wings of swifts and swallows. Fast-flying ducks and waders also have narrow wings to keep down profile drag and their good streamlining reduces the amount of parasite power required. The result is a rather flat U-curve showing that these birds fly fast but economically on their daily journeys to feeding grounds. Shorter, broader wings with wingtip slots are designed for high power at low speed. One example is the catbird of North America whose name derives from its mewing call. It lives in thickets where it makes short, slow and therefore strenuous flights amongst the tangled vegetation.

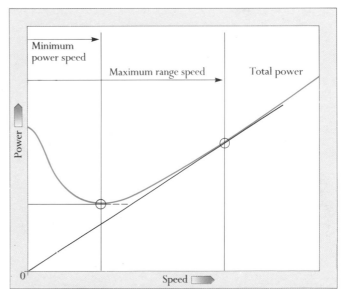

The power needed to fly steadily depends on the bird's speed. At low speeds it expends induced power to sweep the wings through the air. This effort is reduced when the bird's forward movement generates lift, but drag increases at higher speeds.

A bird can vary its speed, but flight is most economical between the minimum power speed and maximum range speed, depending on the object of the flight. There is a speed where power expenditure is minimal, but absolute economy is not always the best strategy.

WHAT SPEED TO FLY

Flying slowly, as at take-off and landing, is strenuous because of the high induced power requirements. Flying fast is also strenuous because drag builds up and parasite power is needed to overcome it. Between these extremes there is a range where flight is economical. A horizontal line drawn at the level of the power that the bird can exert continuously intersects the power-speed curve at the maximum and minimum speeds for steady flight. At the bottom of the U, there is a range of speeds where power requirement is minimal and its flatness shows that the bird can vary its speed in this range with little change in energy consumption. This range covers the bird's normal operational speeds.

Where power requirement is lowest because the wings are generating lift by forward movement but drag has not built up, there is the *minimum power speed*. This is the speed which allows the bird to fly as long as possible on a unit of fuel. It is the speed of a bird that is not in a hurry and needs the greatest endurance to stay airborne for a long time, for instance a harrier or a buzzard searching the countryside for food. In aircraft terms, it is the long-range maritime patrol that must quarter the ocean for as long as possible before going home.

When going on a journey, such as bringing food to its nestlings or migrating to winter quarters, a bird flies faster than the minimum power speed. It is interested in reducing the energy consumed per unit of distance and so flying as far as possible, rather than for as long a time as possible with the minimum power speed. Although this sounds contradictory, a slight increase in power and speed lets the bird fly farther without refuelling. This is the *maximum range speed* and is the speed at which the ratio of power to speed is at a minimum. It is calculated from the power-speed curve by drawing a line from the origin of the graph at the left to strike the curve at a tangent. In aircraft terms, the maximum range speed is the cruising speed which an airline pilot will choose to make a transcontinental flight, for example from Kennedy to Heathrow, as economically as possible.

These curves are based mainly on calculations, so an obvious question is whether they are a true representation

BELOW The Arctic tern's forked tail and pointed wings, which produce lift with little drag, enable it to float slowly over the sea, searching for food with little effort.

ABOVE A kestrel brings a large lizard to its nestlings. It has sufficient spare power in its flight muscles to carry cargoes of food from the hunting grounds to the nest.

BELOW Folding the tail streamlines the Arctic tern for high-speed flight when commuting between nest and feeding ground while busy rearing its young.

of the way birds fly in real life. The power-speed curve shown on page 43 is an idealized form based on aerodynamic theory and laboratory measurements of energy consumption. The details for any species of bird depend on its wing shape and other characteristics. The power-speed curve was originally worked out by Colin Pennycuik using domestic pigeons which were trained to fly in a wind tunnel so that their power output could be measured at different speeds. The pigeon is therefore a good example of how the power-speed curve can be used to illustrate a bird's flight performance. The curve shows that the minimum power speed is about 18 kilometres (11 miles) per hour and its maximum range speed is about 43 kilometres (27 miles) per hour.

For comparison, Colin Pennycuik drew curves for two very different types of bird. Hummingbirds are famous for hovering in front of flowers to sip their nectar. Hovering is very strenuous and is only possible as a lifestyle because nectar is a rich source of energy. The power-speed curve for hummingbirds shows that their level of sustainable power is well above that needed for continuous hovering, whereas the pigeon's curve shows that it can hover only briefly.

At the other end of the size scale, the Californian condor has difficulty with powered flight. It is incapable of generating enough induced power with its wing muscles to take off steeply from the ground like a pigeon and has to run about 20 metres (66 feet) to take off like an aeroplane, generating supplementary lift from the airflow across its wings. Even when airborne it does not have enough power to sustain its minimum power speed. The condor is now extinct in the wild because of persecution. That it was such a magnificent spectacle when it still flew freely over the rugged mountains of southern California is due to its ability to soar on its huge, outstretched wings. Claude Brown has described four condors being thwarted by a sea breeze springing up as they attempted to head west towards the Californian coast. The birds were unable to make headway despite flapping hard. Eventually they found a thermal – warm, rising air – and were able to climb over the top of the localized headwind and glide towards the sea.

Attempts to match the calculated maximum range and minimum power speeds with actual flying speeds of wild birds have been frustrated by the difficulty of measuring the speed of wild birds. Moreover, speed will depend on what a bird is doing, how much weight it is carrying in the form of extra fat for migration or food for its nestlings, and on the speed and direction of the wind. The usual assumption is that a bird cruising on a straight course long enough for a recording to be made, provided that there is no more than a light breeze, is flying near the maximum range speed, which can be matched with the calculated value.

LEFT A barn owl's flight speed depends on circumstances. When hunting, it flies slowly and economically. The sight of prey brings a swift swoop and pounce and the owl will then fly rapidly back to its hungry nestlings with its quarry.

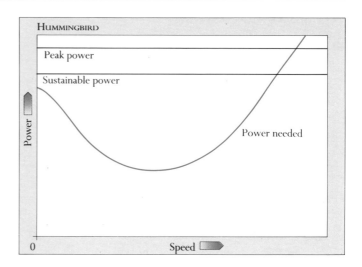

HUMMINGBIRD

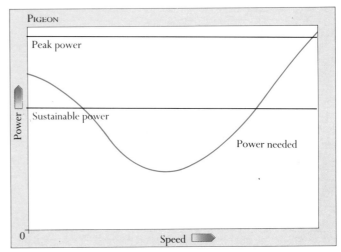

PIGEON

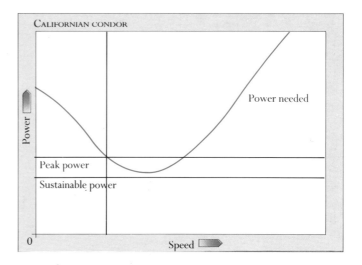

CALIFORNIAN CONDOR

ABOVE Flapping flight becomes harder as birds get larger. Tiny hummingbirds have enough power to hover with ease, while a pigeon takes off by using more power than it can sustain in level flight and can hover only momentarily. A Californian condor can take off only by leaping from a perch or with a helping wind and it remains aloft in still air only with difficulty.

SPEEDS IN KILOMETRES PER HOUR (1.609 km = 1 mile)

Species	Observed speed	Minimum power speed	Maximum range speed
Wandering albatross	54	44	72
Wilson's storm petrel	40	20	35
Fulmar	47	38	63
Grey heron	43	36	50
White stork	67	38	65
Mallard	65	38	68
Kestrel	32	23	40
Common crane	68	46	78
Shag	55	51	83
Herring gull	41	37	61
Puffin	63	42	70
Swift	23	17	24
Longtailed hermit (hummingbird)	43	20	27
House sparrow	35	21	40
Starling	34	28	47
Carrion crow	50	29	52
Swallow	32	16	27

The table on this page gives some reliable figures taken mainly from radar measurements of birds cruising in a light breeze or calm air and some comparisons are made with calculated speeds. They suggest that most birds cruise more slowly than the calculated maximum range speed, but there are some unexpected differences which help to highlight aspects of the birds' flight styles. The wandering albatross and fulmar fly more slowly than expected from calculations but, as we shall see in Chapter 4, they are underpowered and rely more on soaring than flapping flight. As well as seeking economy, they may be flying slowly so that they can scan the sea thoroughly for food. The shag also seems to have muscles that are too small to enable it to reach its calculated cruising speed, but the main surprise is the relatively slow calculated maximum range and minimum power speeds for the swift, which are less than those for the similar-sized house sparrow. Everything about this bird, from its name to its physical form, proclaims a life of speed. As the British poet Ted Hughes wrote in 1976:

> *Erupting across yard stones*
> *Shrapnel-scatter terror. Frog-gapers,*
> *Speedway goggles, international mobsters.*

The narrow, swept-back wings with flat camber and the large flight muscles are usually signs of high-speed flight, yet, according to the ornithologists who have studied swifts, they forage for flying insects at little more than the calculated minimum power speed and migrate at the maximum range speed, neither of which are very fast. There is no doubt that swifts move very fast when chasing each other around the houses, but they are probably in a shallow dive and flapping harder than usual. Obviously a bird that spends its life on the wing must practise economy.

Maximum speeds are even more difficult to measure and some highly suspect claims have been made. Some record books give the laurels to the spine-tailed swift which apparently attained an amazing 320 kilometres (199 miles) per hour. This speed was calculated by timing swifts with a stopwatch as they flew past a bungalow in Assam and disappeared over a ridge of hills 'exactly 2 miles away'. It seems wildly inaccurate to base timings on birds disappearing at such a distance. Perhaps the swifts plunged out of sight before they reached the hills; to make the observation more reliable I would have expected the recorder at least to have had someone stationed on the hills to wave a flag at the instant that the swifts flew over.

The record for the fastest level flight – 76 kilometres (47 miles) per hour, reliably timed by radar – is held by the eider duck and it is unlikely that any bird will be found to fly much faster in a straight line. The peregrine in its headlong stoop is probably the fastest living creature; it has been recorded by radar at 180 kilometres (112 miles) per hour, although a pilot reported being overtaken by a peregrine when he was making a practice attack on a flock of ducks by diving on them at 280 kilometres (174 miles) per hour. Theoretically, a peregrine dropping out of the sky with its wings folded could reach a speed of about 375 kilometres (233 miles) per hour as it finishes its stoop, but it may keep below the maximum possible speed so that it can maintain enough control to track the movements of its prey.

THE ROBIN'S ENERGETIC LIFE

The popular image of the robin is a friendly bird perched on the handle of a garden fork or a low branch, waiting to pounce on worms thrown to the surface as the gardener turns over the soil. The habit started in the primaeval forests of Europe, where robins learned that there was an easy meal to be found by following the movements of other animals: flying down to snap up insects exposed as large mammals scraped the leaf litter with their hooves or worms coming to the surface to escape a burrowing mole.

When a robin is feeding, it does not fly much. It either hops across the ground or keeps watch from a vantage point and flies down and back in short flights lasting less than a second. Compared with a swallow, which is in the air for as much as eight hours of the day hunting airborne insects, the robin spends only a fifth of the day feeding and is airborne for only a few minutes. Nevertheless, a significant part of the robin's energy is used in flight.

Experiments have shown that robins use more energy for these short flights than would be expected from theoretical calculations. This is because they are flying more slowly than the calculated minimum power speed (13.8 kilometres/8.5 miles per hour instead of 21 kilometres/13 miles per hour). They are starting and stopping so quickly that they never get up speed and, as the power-speed curve shows, low speed flight, when a bird is almost hovering, is very expensive. The pay-off is that while other birds fly in search of food, robins do not take off until they have spotted a worthwhile morsel, so they are virtually guaranteed a good return for their effort.

ABOVE A robin brings food to its nestlings. Rearing a family means a heavy burden in extra flight costs.

BELOW Robins usually fly down from a perch to seize a worm or insect, but this one has modified the technique to exploit a very rich source of food.

BUILT FOR FLIGHT

The secret of successful flight lies in the combination of high power output and low weight. In 1680 Giovanni Borelli understood this when he wrote of birds: 'The power of the wings is increased in duplicate ratio: firstly, by the increase in the force of the muscles, and secondly by the decrease of the weight to be supported.' For these twin reasons Borelli realized that humans would never fly like birds, and that 'the Icarian invention is entirely mythical because impossible; for it is not possible to increase a man's pectoral muscles … therefore flapping by the contraction of muscles cannot give out enough power to carry up the heavy body of a human'.

The stamina of birds is amazing. A flying budgerigar has a steady power output of 0.186 watts per gram of muscle, whereas a human sprinter manages only 0.165 watts per gram for a few seconds and is left exhausted. Birds have achieved both increase in power production and saving in weight through far-reaching changes to their anatomy and physiology. It has been estimated that an angel would need a chest 2 metres (6½ feet) deep to house the muscles necessary for flying. No doubt this could be reduced if he saved weight on his other organs – and perhaps there are no legs under those flowing robes. It is a facetious idea perhaps but it draws attention to the radical differences between human and bird bodies.

WEIGHT REDUCTION

The largest bird that ever flew was *Argentavis magnificens* whose prehistoric remains were discovered in Argentina. It was a vulture-like bird with a wingspan estimated at 7.5 metres (25 feet), and its primaries measured 1.5 metres (5 feet). It weighed as much as 77.5 kilograms (171 pounds), which is well above the theoretical limit for flapping flight. It would have had to rely, like modern vultures and condors, on soaring in thermals over the Argentinian savannah and would have been able to take off only with assistance from a stiff breeze.

Argentavis could only have survived in special circumstances. For most birds lightness is a virtue, although weight reduction has had to be achieved without loss in strength because the birds' frame must still withstand the considerable stresses of flying. The advantage of lightening the body is that it leaves power in hand for manoeuvring and carrying food. Reducing body weight also reduces the amount of power needed for flight, so weight-saving has been an essential part of the birds' conquest of the air and reaches its ultimate in the magnificent frigatebird. Designed for soaring over tropical seas, it has a 2.3 metre (7½ foot) wingspan and weighs 1.5 kilograms (3 pounds), yet its skeleton weighs only 100 grams (3½ ounces), which is half the weight of its feathers.

A major theme in the evolution of birds is strength with lightness. The process had barely started at the stage of *Archaeopteryx* and a comparison of its fossils with the skeletons of modern birds shows how much has since been achieved.

The skeleton of a modern bird is reduced to hollow girders and flat plates by the fusing of bones at their joints. The greater rigidity this gives in turn cuts down the weight of muscles and tendons needed to hold the posture of the skeleton. The tail has been reduced from the long balancer of *Archaeopteryx* to a stub (the 'parson's nose') of limited flexibility for mounting the fan of tail feathers. The tail is joined to the backbone which is fused to the pelvis to make a very light but stiff structure. The neck, by contrast, remains extremely flexible so that the beak can be used as a 'universal tool'. The ribs are flat plates with flanges interlocking with their neighbours for extra strength.

The skull is also lightweight and the heavy jaws, chewing muscles and teeth of *Archaeopteryx* have been replaced in modern birds by light jaws covered in the horny beak. However, in place of teeth for chewing food, many birds have a heavy muscular gizzard for grinding coarse food and little, if any, weight is saved. One advantage is that the gizzard, being part of the stomach, is nearer the centre of gravity, which helps to keep the bird stable when flying.

The internal organs have also been subjected to ruthless weight-cutting. Instead of storing heavy urine in a bladder, birds excrete nitrogenous waste in concentrated form as uric acid. The reproductive organs of both sexes atrophy almost to the point of disappearance outside the breeding season. The female has only a single ovary which increases in size by 1,500 times as the breeding season approaches. The need to reduce the reproductive organs is demonstrated by the behaviour of the female sparrowhawk. Before nesting begins, her weight goes up by 13 per cent through the growth of the ovary and the laying down of a fat reserve. This upsets her aerodynamically, increasing profile drag through her stout outline and increasing her wing-loading. Her hunting is impaired and she increasingly relies on food brought by the male. As egg-laying approaches, her flight becomes noticeably laboured.

The birds' retention of the egg-laying habit, in contrast to the live-bearing of mammals and many reptiles, has often been said to be a result of the need to save weight. Although bats bear live young they produce only one at a time; egg-laying overcomes this restriction in breeding rate, so the female bird can have a large family because she is hampered only during the short space of time – about a day – that it takes to manufacture each egg. This may be true but Professor Hans-Rainer Duncker, a German physiologist, points out a more important reason. A chick hatching from the egg takes one to three days to switch from breathing through the egg membranes to the complex system of lungs and airsacs. It is difficult to see how this can take place other than in an egg; live-born animals have to change very quickly from getting their oxygen from the mother's bloodstream to breathing air. Thus the unique breathing system of birds prevents them from giving birth to live offspring.

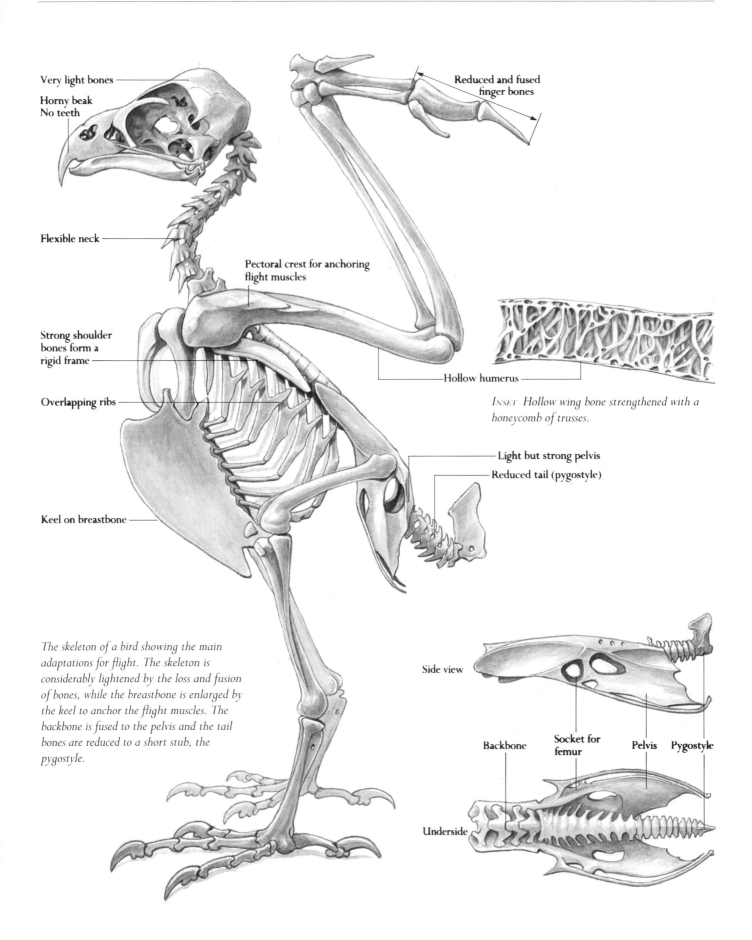

Very light bones

Horny beak
No teeth

Flexible neck

Pectoral crest for anchoring
flight muscles

Strong shoulder
bones form a
rigid frame

Overlapping ribs

Keel on breastbone

Reduced and fused
finger bones

Hollow humerus

*INSET Hollow wing bone strengthened with a
honeycomb of trusses.*

Light but strong pelvis

Reduced tail (pygostyle)

*The skeleton of a bird showing the main
adaptations for flight. The skeleton is
considerably lightened by the loss and fusion
of bones, while the breastbone is enlarged by
the keel to anchor the flight muscles. The
backbone is fused to the pelvis and the tail
bones are reduced to a short stub, the
pygostyle.*

Side view

Backbone

Socket for
femur

Pelvis   Pygostyle

Underside

A diagrammatic representation of the section through a bird's body to show the mechanism of the flight muscles. The muscles are anchored to the breastbone and pull on the humerus (upper arm bone) of the wings.

A: Contraction of the large pectoralis muscles pulls the wings down.
B: Contraction of the smaller supracoracoideus muscles acts through 'pulleys' to pull the wings up.

Saving on superfluous weight extends to the diet. Birds need plentiful, energy-rich food to sustain flight but it must be digested easily and the waste eliminated quickly, so most birds eat flesh, insects, seeds and fruit. Diets of bulky, low calorie leaves and grass are uncommon. The hoatzin is one of the few specialist leaf eaters. It stores leaves in its large crop like a ruminating mammal, but it then has difficulty in flying. Other birds that eat a fair amount of herbage are generally those that do not fly much, such as bustards, quails and pheasants. They have small but frequent meals so their weight never becomes excessive and they can take off if forced to escape from potential danger.

ADAPTATIONS FOR HIGH POWER

The most obvious features of a bird's internal anatomy are its large breast muscles. We see them every time a chicken or turkey is carved. The breast meat is composed of two muscles, one lying over the other. The pectoralis on the outside pulls the wing down and the inner supracoracoideus pulls the wing up by means of a tendon that passes over the top of the shoulder joint, like a rope over a pulley. One advantage of this unique system over the conventional use of the shoulder muscles for raising the forelimb is that it keeps the main muscle mass, and hence the centre of gravity, low in the bird's body. This improves its stability.

On average, the breast muscles of a bird comprise 15 per cent of its total body weight, but powerful fliers such as pigeons have breast muscles weighing as much as a third of the body weight. To be effective this mass of muscle needs a firm anchorage and copious supplies of energy.

The anchorage of the breast muscles is provided by the breastbone or sternum, with its keel or carina, and partly by the wishbone and coracoid. In any animal the area of bone to

which a muscle is attached gives a good indication of the strength of that muscle. The bird's shield-like breastbone, which is already much larger than in a mammal of similar size, is further enhanced by the keel and the whole area for muscle attachment is obviously huge. At the other end of the muscle, a good anchorage on the humerus is provided by a flange called the deltoid or pectoral crest.

To prevent the body collapsing and the organs being compressed as the pectoralis contracts, the skeleton is braced and strengthened. The keel does more than provide extra area; it holds the pectoralis away from the supracoracoideus so that the latter is not squashed against the breastbone when the pectoralis contracts. The shoulder or pectoral girdle forms a rigid frame consisting of the shoulder blades or scapulas, the coracoids (bones which are missing in mammals) and the collarbones or clavicles. At their junction there is a gap, the triosseal (meaning three-boned) canal, which forms the pulley for the tendon of the supracoracoideus muscle. Ostriches, kiwis, rheas, cassowaries and emus, which have given up flying and have only vestigial wings, have lost the wishbone and the keel. They are called ratite birds, from the Latin *ratis*, a raft, because of their flat breastbones. (All other birds are called carinates from their possession of the keel or carina. Some other flightless birds have almost lost the keel, but the penguins retain it because they have large breast muscles for swimming.)

The shoulder blades form the anchorage for the wings and are secured to the ribs by tough ligaments, while the coracoids and clavicles act as struts to hold the shoulder away from the breastbone. In most carinate birds the two clavicles are fused into the wishbone or furcula, but they are separate in many parrots, toucans and barbets although it is not known how this affects their flight.

FUEL FOR FLYING

When flying steadily, a bird uses 10 to 15 times as much energy as it does when quietly perched. If pressed, when taking off or being pursued, it can double this rate. The energy for muscular work comes from the oxidation of fuel, but instead of the explosive oxidation that takes place in an internal combustion engine, muscles oxidize carbohydrate through a long series of biochemical reactions. The end result is the same:

$$\text{Fuel} + \text{oxygen} \rightarrow \text{carbon dioxide} + \text{water} + \text{energy}$$

In the muscles this basic process is called respiration, a term that is often used for the exchange of air in the lungs, which should be called breathing or ventilation. Like all chemical reactions, respiration proceeds more rapidly at higher temperatures, so a high body temperature enables the muscles to work more efficiently. Birds have resting temperatures of around 40 degrees centigrade, which is several degrees higher than in mammals, and when flying their temperatures rise a couple of degrees further.

The main fuels in bird muscles are fat and the carbohydrate glycogen. For regular and sustained production of energy, birds burn fat. Although fat is scarce in the diet of most birds, it is manufactured from carbohydrate in the liver. The transformation is worthwhile because fat can be stored easily. Glucose and other sugars are stored after transformation into glycogen by the addition of water, so carbohydrate is a bulky energy store, whereas fats are stored 'dry' in masses under the skin and around the internal organs. Another advantage of fat

is that it yields twice as much energy as carbohydrate so, combining the two advantages, stored fat contains eight times as much energy as stored carbohydrate.

Glycogen does have one advantage: it can be metabolized anaerobically without oxygen, to provide bursts of high power. Human sprinters respire anaerobically because their lungs cannot supply enough oxygen to the muscles during extreme exertion. The glycogen is transformed into lactic acid which is later transformed into carbon dioxide and water when the sprinters pant after the race. Lactic acid is a poison and its accumulation in the muscles limits stamina. Birds use glycogen to respire anaerobically at take-off and when flying fast, but as the flight steadies to cruising speed they switch to burning fat. They also change the way they use their muscles.

Muscles are composed of two types of fibre which are represented by the 'light' and 'dark' meat of the roast chicken. 'Dark' meat is darker red in the raw state because of the many capillary blood vessels running through it and the higher concentration of the red, oxygen-carrying pigment called myoglobin. The two types of muscle use different fuels and have different functions. Red muscle burns fat aerobically and is used for 'cruising', while white muscle burns glycogen anaerobically and is used in short-term 'sprint' conditions. So strong, steady fliers have breast muscles dominated by red muscle, while gamebirds (and chickens and turkeys) that take-off and fly short distances when flushed have breasts composed mainly of white muscle. The pigeon, which has an explosive take-off followed by fast, steady flight, has a mixture of both muscle types in its breast.

ABOVE The heavy-bodied ground hornbill is propelled into the air by flight muscles adapted for delivering sudden bursts of power, but sustained flight is difficult.

LEFT A buzzard's economical flight lets it spend many hours in the air while searching for food and defending its territory.

HEART AND LUNGS

For muscles to work well they need an efficient blood circulation system to supply fuel and oxygen and remove waste products. Although there are differences in the blood circulation systems of birds and mammals which reflect their separate lines of descent from reptiles, the mechanisms are essentially the same. Both have four-chambered hearts which direct blood to the lungs and, on its return with oxygen, send it to the muscles and other organs where the oxygen is used in respiration. Carbon dioxide is picked up at the same time and carried, via the heart again, to the lungs for disposal.

One difference between the systems, which relates to flight,

BELOW The breathing system of a bird, showing lungs and airsacs. Air is inhaled through the windpipe (trachea) and passes to the airsacs before flowing through the lungs. The thin-walled airsacs act solely as reservoirs; some penetrate the hollow bones.

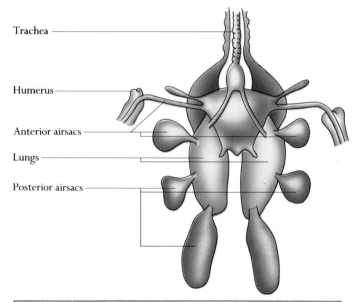

Trachea

Humerus

Anterior airsacs

Lungs

Posterior airsacs

is the larger size and greater power of the bird heart. A sparrow's heart is nearly three times heavier, compared to body weight, than that of a mouse. Small birds have relatively larger hearts than big birds, and birds that make long flights have larger hearts than non-migratory species. Heartbeat is also much faster than in mammals, as you can tell if you hold a bird in your hand. It ranges from a resting pulse rate of 93 per minute in turkeys, 190 in mallards, 700 in chaffinches and 1,000 in hummingbirds, compared with 70 in humans. The combination of large heart and fast pulse rate results in a pumping capacity several times greater than in mammals.

Contrasting with the similar hearts, the breathing system of birds is unique and very different from the mammals' simple lungs working like a pair of bellows. Bird lungs are relatively small, but they are connected to a system of inflatable airsacs that spread through the body and even penetrate the bones and breast muscles.

To prove that the distant airsacs connect with the lungs, the eighteenth-century surgeon John Hunter cut a hole in the humerus of a live chicken's wing then tied up its windpipe to show that it could still breathe. The reason for this unique system remained a puzzle for many years but attracted a variety of often improbable theories. It was said that the airsacs made the bird lighter, thus aiding flight, but this is true only where the airsacs penetrate the bones and, by keeping the core hollow, reduce their density. Even recent books make the surprising statement that hollow bones will allow the bird to carry more oxygen, but why store a commodity that is in ample supply all around?

It now seems to be agreed by anatomists and physiologists

INHALATION

Anterior airsacs   Posterior airsacs

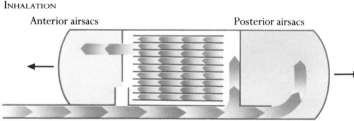

EXHALATION

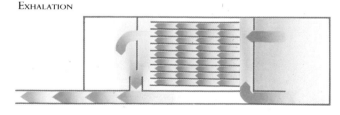

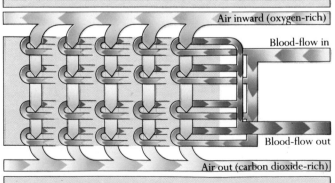

Air inward (oxygen-rich)

Blood-flow in

Blood-flow out

Air out (carbon dioxide-rich)

ABOVE The arrangement of capillary blood vessels wrapped closely around the fine passages in the lungs ensures an efficent exchange of gases. As the air flows through the passages it absorbs oxygen and loses carbon dioxide and water vapour.

ABOVE Diagrammatic representation, greatly simplified, of the airflow through the lungs and airsacs. Movement of the chest expands and contracts the airsacs, working them like bellows, to drive air through the system.

that the function of the airsacs is to provide an efficient exchange of oxygen and carbon dioxide with the bloodstream by maintaining a one-way flow of air through the lungs. The highly stylized diagrams on the page opposite show the flow through the lungs and airsacs. When the bird inhales, air enters the posterior airsacs and some flows through the lungs to the anterior airsacs. On exhalation, air from the posterior airsacs flows into the lungs and the air which had previously filled the anterior airsacs is exhaled. This creates a constant passage of air through the lungs and ensures a more efficient absorption of oxygen and removal of carbon dioxide compared with the mammal lung where only a fraction of the air is flushed out at each breath.

Bird lungs are not worked by a diaphragm; the airsacs are pumped by movements of the hinged ribs and recent research has shown that they are aided by the furcula. The popular names of wishbone and merrythought describe one of the furcula's magical qualities, but it was also used for weather forecasting. Dr Hartlieb, a German physician writing in 1455, asked an army officer what sort of weather the winter would bring. 'This valiant man, this Christian Captain, drew forth out of his doublet that heretical object of superstition, the goose-bone [wishbone], and showed me that after Candlemas an exceeding severe frost should occur.' The more rational function of the furcula is that it helps pump air around the bird's body. In most birds which have been studied (pigeons and crows are exceptions) wingbeats are much faster than the breathing rate (three breaths to 12 to 16 wingbeats in the starling). The furcula acts as a springy coupling between wings and airsacs and, it has been suggested, transmits the force of the wingbeats to pump the airsacs and increase ventilation during flight.

The airsacs seem to be an unnecessarily complicated breathing system because bats fly very well with the standard mammalian in-and-out lung system, but the advantage of airsacs appears to be that birds can fly in the rarified air of high altitude. Whereas mice in a test chamber were barely able to crawl at the equivalent of 6,100 metres (20,000 feet), sparrows could still fly and gain altitude. The one-way system may, therefore, be the mechanism that lets birds fly high enough to cross mountains and migrate in high-speed jet streams. However, not every bird is so well adapted. A starling flying in a test chamber at the equivalent of 3,500 metres (11,500 feet), which is not a great height for a migrating songbird, soon became distressed but recovered when given some extra oxygen.

Most bird flight takes place within 100 metres (330 feet) of the ground, but greater altitudes are reached on migration. The bar-headed goose which migrates over the Himalayas and the Canada goose which migrates over the Rockies encounter falls in air pressure that would incapacitate a mammal. The altitude record is held by a Rüppell's vulture which was sucked into an airliner's engine at over 11,000 metres (36,000 feet). It was probably soaring, so its oxygen requirements would have been much less than in active flight.

FLIGHT MAINTENANCE

A bird sets aside time every day to look after its feathers, because flight becomes laboured and inefficient if the aerodynamic surfaces of the wings and tail are damaged. Odd moments are snatched for preening, but there are also prolonged bouts when the bird gives its plumage a thorough working-over. It gently nibbles a feather or bunch of feathers and smooths them with the edge of its bill to 'zip up' parted barbs and remove dirt or parasites. As part of the preening process, most birds smear their feathers with oil from the preen gland under the tail. At one time it was believed that the oil helped waterproof the feathers or was transformed into vitamin D by sunlight. The evidence is that it keeps the feathers supple and kills bacteria and fungi.

A preening bout is often preceded by a wash. Birds differ in their fondness for bathing. Some species visit garden birdbaths and ponds regularly; others rarely if ever take a dip, but they may get a sufficient soaking from the rain. Pigeons rain-bathe

A trumpeter swan throws its body into contortions as it carries out the essential task of preening.

ABOVE A crimson rosella bathes among waterlilies. Its feathers are fluffed out to let the water penetrate.

LEFT A barn owl draws a flight feather through its bill to mend splits in the vane by rehooking the barbicels.

RIGHT Turkey vultures sunbathing on giant cacti. The sun's rays are believed to improve the condition of the feathers.

as well as taking the traditional bath, and such diverse birds as woodpeckers and parrots wash in the rain, while warblers and hornbills are among the birds that bathe by flapping in wet foliage. Swifts, swallows and kingfishers, even hummingbirds and owls, bathe by dipping into water while in flight.

Although bathing sometimes results in a good soaking, the bird makes sure that it does not get too waterlogged to fly. One function of the wetting seems to be to make the feathers easier to preen and helps spread preen oil over the feathers.

Sunbathing also appears to have a function in feather care. The ornithologist David Houston has shown that twisted vulture feathers straighten out in three to four minutes in the sun but take two to three hours in the shade. When vultures, bateleur eagles, pelicans and storks soar on outstretched wings their long wing feathers become bent. These birds sunbathe with their wings spread, while large flapping birds such as swans and herons do not.

MOULTING

A feather lost through accident is immediately replaced, but all birds replace the entire plumage at intervals. This is necessary because feathers become threadbare and frayed through natural wear and tear. Friction caused by feathers rubbing against each other, abrasion caused by foliage or the fabric of the nest, and even the continual surges of air flowing around the feathers during flight take their toll of the material of the vane. The barbicels and barbules disintegrate, so the cohesion of the vane is lost and it eventually wears away.

Moulting requires an investment in raw materials for the new feathers – equivalent to one quarter of the protein content in a small bird – and an expenditure of energy is necessary for their manufacture as well as for keeping warm and flying while there are gaps in the plumage. Flight must be more difficult during moulting, especially when the primaries are regrowing and the wings are shortened. Large, soaring birds, such as albatrosses and eagles that spend many hours in the air every day, moult slowly so that very few feathers are missing at any time and a complete moult may take two years, by which time the next will have already started.

The extra materials and energy needed for the moult are supplied by eating more and do not usually involve a strain on the bird's resources, but falconers have long been familiar with 'hunger traces' or 'fault bars' in the wing and tail feathers of their charges. Even a brief shortage of food leaves a line of weakness across a growing feather, where the shaft is pinched and the vane lacks barbules. The feather can easily snap at this line and spoil the bird's flight performance. Falconers mend or 'imp out' a weak feather by grafting on a length of feather cut from a store of old ones kept for the purpose. Shakespeare used this expression in Richard II: 'Imp out our drooping country's broken wing.'

There is, nevertheless, a general rule that birds time their moult so that it does not coincide with times of food shortage or important, energy-demanding events such as breeding and migration. The moult of an average small bird takes three or four months from start to finish. It usually takes place after breeding, although some birds start to moult before nesting has finished.

There are other variations in the moulting schedule, such as the European swift which moults after it has reached its

A ptarmigan changes colour by moulting as the snow melts. The patchy plumage, with feathers missing, hinders flight.

winter quarters in Africa and the female sparrowhawk which starts to moult soon after laying her eggs. She relies on her mate for food, so she takes the opportunity to replace her feathers while she does not have to hunt. She can still flap heavily away if disturbed, but some female hornbills, which are walled into their nests and have no opportunity for flight, shed all their wing and tail feathers together.

Such a catastrophic moult is not unusual. Ducks, geese, cranes, rails, and many auks and divers survive a flightless period during the moult by retiring to the safety of open water or dense vegetation where they replace their feathers quickly. Barnacle geese choose to moult in richly vegetated lakeside areas in the Arctic tundra, so they can feed well and take refuge from Arctic foxes on the water. Their entire moult lasts little more than five weeks, but the geese are grounded for only about 25 days because they can fly before the flight feathers are fully grown.

The reason for these catastrophic moults is not always clear, but the larger species of auks, such as guillemots and razorbills, have such high wing-loadings that the loss of only a few feathers would make flight too strenuous. Smaller species, such as the little auk, have lower wing-loadings and moult in

A cape pigeon gliding overhead shows missing inner primaries caused by moulting. Feathers are lost symmetrically on each wing.

the normal way without losing their power of flight.

The way that moulting fits into the lifestyle of a bird is neatly shown by the waders that migrate to the Arctic to breed. They nest during the two to four months of the Arctic summer, then migrate thousands of kilometres south for the winter. Dunlins nesting around Point Barrow in Alaska start to moult their primaries as soon as the eggs are laid, which gives them 60 days to complete the last feather before they depart at the end of summer. The autumnal voyage south is leisurely compared with the race north in spring and those birds that have not completed their wing moult do so on the way. Moulting of the body feathers is less important. It not only starts later but will be delayed in a poor summer, while changing the flight feathers has to be completed at any cost.

One thousand kilometres (620 miles) farther south in the Yukon, summer lasts two months longer and the dunlins do not start moulting until the chicks are growing up; they change their primary feathers in a leisurely fashion in the 90 days before heading south. In the Old World, the story is very different. Some dunlins start to moult on the breeding grounds, then stop until they get to Africa. Others moult at a staging post, such as the extensive mudflats of the Waddenzee or the Wash. Yet others do not even start until they have got to winter quarters where, with plenty of food and no need to fly far, they moult over a period of around 150 days.

CHAPTER 3

FLYING SKILLS

A bird needs special skills to take off and land, as well as to control its flight. The problem of take-off is to produce enough power to obtain the lift for getting off the ground. When landing, the bird has to maintain lift so that it can slow down without falling out of the air. At all times when it is in the air, it has to maintain stability and to undertake the manoeuvres that enable it to make the best use of its powers of flight.

A wandering albatross shows off its mastery of flight.

TAKING OFF

An aircraft takes off by taxying down a runway until it reaches flying speed, the speed at which the airflow over its wings generates enough lift to support it. A bird has an advantage over the aeroplane because it can leap into the air and flap its wings to generate lift, which enables it to fly away while its airspeed is still well below stalling point. Many birds take off without any forward movement; all the lift is generated by the wingbeats. This is no problem for small birds with plenty of spare power, but larger birds employ a special form of wingbeat for taking off and when flying slowly at other times. It generates large amounts of lift with little forward speed but the drawback is that it requires massive exertion, like an athlete's sprint start that cannot be sustained for the whole race. Once the bird is fully airborne, its wingbeats slow down and become shallower until it falls into the pattern of straight and level flight described in Chapter 2.

Disturb a pigeon when it is on the ground and it will clatter heavily into the air, climbing steeply for 20 to 30 metres (65 to 100 feet). The pattern of the first few wingbeats at take-off has been revealed by photographing domestic pigeons with cine cameras running at 500 frames per second. The pigeon opens its wings and straightens them over its back, then it jumps up so that it is far enough off the ground to make a full downstroke and already gaining forward speed. Next, the wings swing downwards and forwards so that the tips finish by sweeping almost horizontally, scything through the air to create an airstream that will generate enough lift to maintain the upward momentum.

During the next few wingbeats the pigeon rises almost vertically into the air. The two aspects obvious to the naked eye are that the body is almost upright and the wingbeats are much deeper than normal (sweeping through an angle of 280 degrees, the greatest that is anatomically possible). The

bending of the primaries shows that there is a strong lifting force on the outer wing but, without much forward speed, the inner wing plays a very minor role.

With these slow wingbeats the pigeon would drop back between the downstrokes were it not for the 'hand', bearing the primary feathers, turning over to give extra lift. At the start of the upstroke, when the wings are almost pointing forward like a springboard diver's arms, the wing partly folds and the wrist rotates through almost 180 degrees, so that the primaries turn upside down and open like a venetian blind with their leading edges pointing in the direction that they will be brought on the upstroke. Resistance to the backward and upward movement is very effectively reduced, but the orientation of the primaries has a greater significance. As the wing rises and begins to spread again, it is flicked smartly upwards and backwards. The resulting airflow strikes the inverted primaries and is converted into lift and drag. As the wing is moving backwards, the drag, in fact, propels the bird forwards. So, surprising as it may seem, the main lift and propulsion in this type of wingbeat comes on the upstroke.

Pigeons are probably helped by a so-called 'clap-and-fling' technique which creates lift by allowing air to flow into the space opened up between the wings. The characteristic clapping of their wings as they take off is caused by the wings meeting over the back and being flung open like a book. The leading edges are whipped apart, while the rear edges of the flight feathers are first pressed together and then peeled apart,

RIGHT When a wood pigeon takes off, its primaries turn upside down and separate on the upstroke.

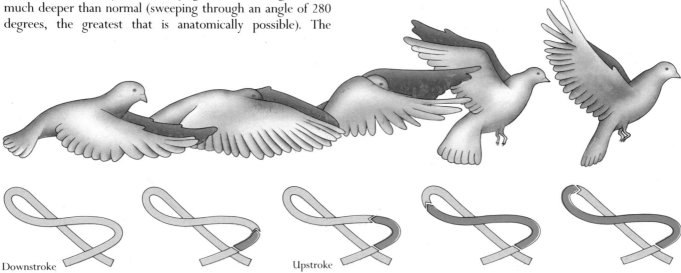

ABOVE The wingbeat cycle of a pigeon as it takes off. The wingtips move in an approximate figure-of-eight. Left to right: the downstroke is nearly complete, with the wingtips sweeping forward. On the upstroke, the wings fold and the primaries open as they are brought up and back. Then the outer wing twists and flicks upwards to provide additional lift.

Downstroke

Upstroke

A multi-image photograph shows a robin taking off when alarmed. One eighth of a second elapsed between the first and last image. The robin is already turning and flying away rapidly.

A brilliantly coloured harlequin duck runs across the water to get up flying speed. Propulsion by the legs reduces the effort needed from the wings to lift the duck into the air.

sucking air explosively over the wings to generate lift. It is not yet known whether any other birds employ the clap-and-fling technique, but it is quite possible that pheasants and other heavy-bodied species use a similar method for their vertical take-off.

This type of take-off is very tiring and demands a margin of power well in excess of that required for level flight. The curve of power required to fly at various speeds shows that a pigeon has plenty of power to fly slowly, below its minimum flying speed, for a short time (i.e. the curve is below the line of the pigeon's maximum power output but above the line for the maximum continuous power output). This means that although it can take off by jumping into the air, the effort is so great that the pigeon cannot take off from the ground twice in two minutes. If it is frightened into the air it either takes refuge in a tree, where another take-off can be achieved simply by dropping off the branch, or it flies well away from the source of disturbance.

As birds increase in size they generally have less power to spare and rely more on forward speed to supplement the lift from their wingbeats. Birds as large as condors have barely enough power to sustain level flight and the ornithologist

J. McGahan tells of flushing an Andean condor from a beach on a calm day. It tried to escape over a ridge but landed below the crest and ran up it. In 20 seconds of flight it had achieved a height of only 14 metres (46 feet). If it had not been in danger it would not have attempted to take off.

Bustards, pheasants and similar ground-dwelling birds are heavy-bodied, having reversed some of the sacrifices that birds have made to become efficient fliers. However, they need to take off quickly when chased and their short wings give them rapid, deep and powerful wingbeats from the moment of take-off, while wingtip slots reduce drag or give extra lift at low airspeeds (see page 36).

The great bustard can take off with a short run of four or five paces if startled into an emergency retreat, but the effort is extreme and it prefers a longer run to get up speed before lifting off. The power margin is so low that it cannot take off in wet weather until its wings have shaken off excessive water; thus when dogs were used for pursuing bustards in seventeenth-century England, it was a good ruse to go hunting on frosty mornings when the bustards were iced up and grounded. The larger kori bustard has even greater difficulty. At maximum power it is like a condor; it can just take off after a long run, but it can barely stay airborne long enough to fly clear of pursuers.

In view of these problems it is very odd to find another bustard, the lesser florican, albeit a small species, employing vertical take-offs in its display flights. The male flies perpendicularly out of the long grass to a height of about 10 metres (33 feet), then folds its wings and drops back to the ground. It may make 30 or 40 of these strenuous 'flying leaps' in the space of half an hour and over the course of a day the energy expenditure must be enormous. The surprising aspect is that slow-motion film shows the florican rising with its body horizontal, instead of tilted well back like pigeons or pheasants taking off, and its wings flap up and down rather than sweeping forward and backward like the pigeon shown on page 62. It is difficult to see how this action can create the airflow over the wings to give the florican enough lift, but the deeply emarginated primaries of this species may be of assistance.

The ability to take off and climb steeply is related to the size of the supracoracoideus muscle that raises the wing (see page 52). In large, slow-flapping birds and in many small species the supracoracoideus weighs less than one tenth of the pectoralis, but it is one quarter the weight of the pectoralis in pigeons and one third in some gamebirds.

The skylark, incidentally, has a large supracoracoideus which may be important for its song-flight:

> *And singing still dost soar, and soaring ever singest.*

Shelley's *Ode to a Skylark* is more inspirational than scientifically accurate. With its tail spread, the male skylark ascends to a height of 100 metres (330 feet) in a tight spiral, heading into the wind for extra lift, then hovers for a minute or more before gliding steeply back to earth. The excellent rate of climb is also an advantage when pursued by a merlin because the lark can escape upwards, leaving the merlin climbing slowly until it gives up.

ABOVE A little owl caught in the act of pouncing on prey. It delays opening its wings until it is clear of obstructions and will then hurtle down and strike before the victim is alerted.

RIGHT The long wings and short legs of a swift are signs of its supreme aerial capabilities, but they make take-off difficult unless the swift launches itself from a perch.

The difficulty of climbing vertically is a cause for disaster among birds that get trapped in chimneys. Although often attracted to such places for nesting, jackdaws, which have a small supracoracoideus, frequently find their only exit is down through the fireplace. There is no problem for chimney swifts which nest in chimneys and at the bottom of hollow trees, or other species of swifts that nest up to 70 metres (230 feet) underground in potholes. Swifts have very large supracoracoideus muscles and, although the common European swift does not hover, this is a feature of other species.

The problem of weight is compounded by the shape of the bird. Many otherwise graceful and adept fliers make an ungainly transition from surface to air because they cannot flap vigorously for lift generation without smashing their long wings on the ground. The situation is made worse if they have

LEFT The mallard has enough power to launch itself straight up from the surface of the water. Its wings sweep through the air to lift it in an almost vertical climb.

BELOW Where black-browed albatrosses nest on flat ground, they need runways so they can take off by taxying along for several metres.

short legs. Swifts, for example, like to settle on trees or walls rather than on flat ground because of their tiny legs and very long wings. However, contrary to common belief, a grounded swift can take off if it is disturbed: it raises its wings over its back and, with one sharp downstroke, lifts into the air and flies away.

The problem is most acute for waterbirds with their high-aspect-ratio wings and short, but powerful, legs set well back on the body for efficient swimming. Divers, which have very high wing-loadings and tiny alulas, take off only with a long run and slow climb. Albatrosses and their relatives are nearly as bad, but they are helped if they can taxy into the wind and catch updraughts from waves or cliffs. When I was catching albatrosses on South Georgia to ring them, I learned the trick of intercepting them as they took off into the wind. If there was no more than a light breeze, take-off required a long run and I got my bird, but a strong wind often helped it escape. The frigatebirds have the greatest problem with their 2-metre (6½-foot) wingspans and minute legs that are no use for running, so they cannot take off from flat land or water. They only come to rest on trees or the guy wires of radio masts where they can take off again by dropping off the perch.

LANDING

The skill with which birds make an accurate and graceful landing disguises the problems that face them. A bird must reduce its speed, eventually below its minimum flying speed, so that it can touch down without damaging itself. This is difficult for birds with high wing-loadings that cannot fly slowly, so they tend to land heavily and run to a standstill. As with taking off, the bird has to produce sufficient lift to support itself while flying slowly but, whereas a bird taking off is launching itself into open space and the freedom to manoeuvre, on landing it is often heading into a tight space and danger.

Small birds usually land without difficulty. They simply pitch in and come to a halt with a flurry of wings. A very awkward perch may present problems, but many birds, such as tits and warblers which forage amongst dense foliage, are adept at landing upside down. With larger birds – pigeons or even starlings – there is a more complicated landing sequence.

APPROACH

An aeroplane starts the approach phase of its landing sequence when it loses height and lines up on the runway. For the passengers, the first sign that they are about to land is the lowering of the flaps which, in effect, increase the angle of attack and create drag. This is the equivalent of a bird tilting back its wings, especially the inner wings, until they almost stall, as can be seen by the covert feathers lifting. A wing near stalling gives high lift as well as drag, which helps to keep the bird airborne while it is slowing down. Complete stalling is prevented by the outer wing twisting down to lower its angle of attack, while the alula opens and the primaries spread to maintain the smooth airflow over the aerofoil.

The art of landing is to touch down at the correct spot (accuracy is particularly important when landing on a perch) at the right speed. If an aeroplane is undershooting and the pilot puts its nose up, speed falls off and it may stall; if it is overshooting and the nose is put down, speed increases and the aeroplane may land heavily. Birds avoid this problem by varying the set of their wings to alter their relative lift and

RIGHT Wings out to brake and feet forward to grasp the perch, a cockatiel makes a delicate landing.

BELOW Despite its long legs and ungainly appearance, a painted stork makes a carefully controlled landing on a slender branch.

drag and you can see medium-sized birds, such as pigeons, thrushes and starlings, adjusting their approach path by altering their airspeed and sinking speed in a kind of 'roller-coaster' flight. Depending on the circumstances, they alternately fold their wings and drop, then open them and level out. Reducing the wingspan increases the glide angle, sending the bird into a dive, but it also increases forward speed which is undesirable on landing, so opening and closing the wings is a technique to achieve a controlled descent. Alternatively, the wings are tilted up to increase drag and check the bird momentarily. Either way, the manoeuvre brings the bird down to a spot which it appears to have chosen well in advance.

Sometimes birds need to land in a hurry. From my study window I have seen magpies dropping out of a tree to pick up food on the ground. The magpie plummets vertically, beak first, with wings folded, until less than a metre from the ground. Then it spreads its wings, fans its long tail, brings its head up and drops on its feet. These are the fastest landings I have witnessed; the magpie's object is to reach the food before another bird takes it. Similarly, in the tropics, vultures hurtle down to feed on a dead animal before other scavengers arrive. They lower their legs to act as airbrakes, trebling the body's drag without creating any lift, so that the sinking speed increases and they drop out of the sky.

Geese need to drop rapidly from cruising height to the ground or water to avoid running the gauntlet of predators or guns. They have a trick called whiffling in which the flock breaks ranks as it approaches the landing place; the geese hurtle earthwards, rolling to left and right, tumbling and

Above Landing at speed is much easier on water. A bufflehead descends quickly but touches down by sliding to a halt using its webbed feet as water-skis.

Left High stalling speed makes landing difficult for a puffin. It is not easy to fly slowly without losing control, but it uses its webbed feet as flaps for extra lift.

turning upside down, to the sound of air rushing through their feathers, finally braking at the last moment to land in an orderly fashion.

Whiffling seems to be like the sideslip landing used by pilots of light aircraft if they are approaching the runway from too great a height. A sideslip reduces lift, allowing the plane to drop without gaining too much speed. Such a technique could be useful both for geese, which have a high wing-loading and an already fast and rather uncontrolled landing, and for the white-collared swift of Central and South America, which nests in caves behind waterfalls. Bat falcons, peregrines and other fast predators gather to catch the swifts as they return in the evening, but the swifts evade them by diving at immense speed, with wings spread for precision steering, and whiffling to enable them to drop more freely.

FINALS

Just before touchdown an aeroplane lowers its undercarriage, throttles back and makes its final preparations for landing. In birds the final stage may include a phase of gliding with the wings and body tilted back in an attitude known to pilots as 'flaring', as seen in delta-winged aircraft such as Concorde. Finally, there is a flurry of exaggerated wingbeats, giving the impression that the bird is back-pedalling, and forward speed drops markedly. In slow motion, one can see that the actions

ABOVE A Canada goose at the moment of touching down. The wings have a very high angle of attack to act as brakes, but some lift is maintained by the open primaries and alula.

are similar to those of take-off, in that the bird's body is almost vertical and the wings are swept to and fro. However, the inner wings have a very high angle of attack and the outer wings are turned almost horizontal so the sweep of the wings generates lift. The result is that the bird slows until it is almost hovering, then gently pitches in with the minimum force.

At the last moment the bird's legs are thrown forward to absorb the impact of landing on the ground. Landing on water is easier. Waterbirds do not have to reduce their speed so much as they can splash down and slide to a halt using their outstretched webbed feet as combined water-skis and brakes. Despite their weight and fast landing speed, swans maintain their grace when landing on water but are most uncomfortable when alighting on land. Small swans – hoopers and Bewick's – regularly land on the ground, often with a short run to bring them to a halt, but the larger mute swan rarely does so. It sometimes mistakes a road for a river and finds itself running hard when it touches down on the unexpectedly firm surface. The ease of landing depends on windspeed. Compared with the heavy touchdown seen in calm weather, swans landing in a strong wind are wonderfully graceful.

ABOVE A goose 'whiffles', banking steeply to lose height quickly.

LEFT Setting down on land requires care. This trumpeter swan has braked as much as possible and will start running as soon as it lands.

Approaching downwind in a shallow glide, they turn into the wind and drop almost vertically, then ease forward to water-ski, still without flapping.

Some waterbirds cannot use their feet as water-skis. Divers, for example, have feet set too far back on their bodies. They sweep down in a steep, fast glide, then level out so that they touch down on their breasts, finally sliding to a halt in a shower of water. Coots have a different problem. They land at speed like other waterbirds but because their feet are lobed rather than webbed, they run over the surface instead of sliding. As they lose lift and speed, their legs gradually sink into the water until the body is afloat.

Long-winged fulmars and albatrosses cannot back-pedal without damaging their ultra-long slender wings. Their method of slowing down without losing too much lift is to 'waggle' their wings as they approach touchdown, tilting them rapidly to and fro so that the angle of attack is raised and lowered. This is not a very effective braking system and they land in a rush when they return to their island colonies in calm weather. The albatross has no option but to approach at its best gliding speed, slowing down at the last moment by waggling its wings and lowering its feet. Its momentum then throws it forward onto its breast while its wings are arched to avoid damage. But if there is a wind to provide extra lift, it glides slowly forwards with its legs lowered and lands with its dignity unimpaired.

Another approach to an awkward landing is used mainly by birds which cannot glide well. Guillemots nest on narrow cliff ledges and require precision landing to avoid collision with the back wall or other guillemots. They approach the cliff below the level of the ledge, then fly up sharply so that gravity cuts

their airspeed. A flurry of back-pedalling controls the final approach; the guillemot hovers momentarily over the ledge then drops onto its feet. This 'ballistic approach' is used by many other birds when landing on a perch and some birds use it for landing on the ground. If there is a good wind, rooks glide close to the ground towards the chosen spot, rear up with wings and tail fanned, then drop on their feet.

A blue-footed booby, a tropical gannet, lands at the edge of its colony. Heading into the wind makes landing easier, but the feet will have to absorb the shock of impact.

CONTROL OF FLIGHT

There are two aspects to the control of flight. A bird has to maintain stability so that it keeps to a straight and level course despite buffeting by the wind and it has to manoeuvre by leaving its level course and turning, climbing or diving. It is easier to understand the problems of flight control in birds by

Landing upside down on the rim of a fat-filled coconut shell requires superlative control, as shown by this nuthatch.

first considering a fixed-wing aircraft. The aircraft is basically stable, which means that if it deviates from straight and level flight forces act on its wings and tail to bring it automatically back on course. It is said to be 'trimmed'. A dramatic demonstration of the innate stability of aircraft was given in 1989 when a pilot baled out of his MiG fighter over Poland and the empty plane continued in a straight line until it ran out of fuel and crashed in Belgium.

Stability requires control in three axes. Pitch is nose-up, nose-down. If the nose drops, a downward force develops on the tailplane, pushing the nose up again until the force cancels out. Roll is the rotation around the long axis. If one wing drops, its lift increases, while the opposite wing loses lift. Yawing is a lateral swing and is brought under control by the tailfin. Manoeuvres such as turning, banking, climbing and diving are effected by the pilot overcoming the natural stability of the aircraft and imposing forces on the wings and tail that will push it in the required direction.

Trimming their craft to fly in a straight line without the constant attention of the pilot was the most difficult problem facing the pioneers of flying and the diaries of Wilbur Wright show that he spent a lot of time watching birds to try to find the solution. In 1900 he wrote: 'My observations of the flight of buzzards leads me to believe they regain their lateral balance ... by a torsion of the tips of the wings. If the rear of the right wing tip is twisted upward and the left downward, the bird ... instantly begins to turn [roll].' Later he noted: 'I think the bird ... retains its lateral equilibrium, partly by presenting its two wings at different angles to the wind, and partly by drawing in one wing, thus reducing its area.' Such

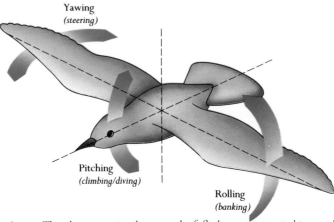

ABOVE The three axes in the control of flight: yawing, pitching and rolling. The equivalent manoeuvres are shown in brackets on the diagram.

BELOW Gliding birds such as harriers and kites may hold their wings in a shallow V. This is called dihedral and corrects rolling. If the bird rolls one way, the lower wing generates more lift, while the upper wing generates less lift.

ABOVE Magnificent frigatebirds have negative dihedral — the wingtips are angled down. This makes the birds unstable but gives them much greater manoeuvrability when chasing other birds.

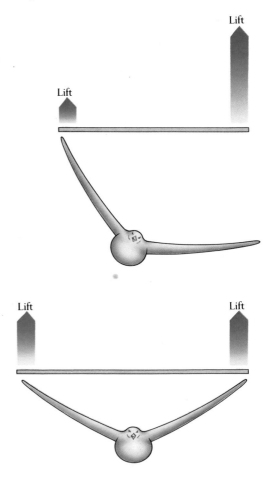

ABOVE The result is that the bird rolls back onto an even keel and stabilizes as the lift of each wing is balanced. Dihedral is commonly seen when pigeons fly down to land.

observations were used by the Wright brothers to develop a system of warping or twisting the wings for flight control.

For the Wrights, copying birds was a blind alley because a bird is essentially unstable. Stability might seem to be sensible but birds have evolved in the opposite direction. The long tail of *Archaeopteryx* gave it stability – an outrigger or keel performs the same function on a boat – but it was aerodynamically inefficient because of its high drag and it hindered manoeuvring. The trade-off in losing stability is greater manoeuvrability, because it is easier to direct an unstable bird or plane onto a new course. The modern bird maintains its flight by small, continuous movements of the wings and tail, as both Wilbur Wright and Leonardo da Vinci noted, directed from a sophisticated control centre in the brain. The bird is like a cyclist who keeps upright by unconsciously moving the handlebars and shifting his weight to steer and balance. The Wright brothers lost their lead to European pioneers who made their aircraft stable on the lines of *Archaeopteryx* and therefore easier to fly. Ironically, the development of

Sparrowhawks exercise great manoeuvrability as they hunt by threading their way through gaps in vegetation at high speed.

computers, which can do the job of the bird's brain, will allow the next generation of fighter planes to imitate birds by abandoning the innate stability of conventional aircraft in favour of greater manoeuvrability.

Some built-in stability has been retained by birds, however. Their bodies hang under the wings and act like the ballast of a ship to keep an even keel and some species hold their wings in a 'V' when gliding. This 'dihedral' configuration stabilizes rolling. Raised wings generate less lift so, if the bird tilts to one side, the lower wing levels out, generates more lift and is pulled up until its lift balances that of the other wing and the bird is level. Dihedral is used by birds such as pigeons as they glide down to land, and by harriers and kites which need to be stable as they glide low over the ground searching for prey. Frigatebirds, on the other hand, need to be unstable for increased manoeuvrability. They use 'negative dihedral', with the wings angled down, which enables them to make incredibly rapid changes in direction as they chase other birds to steal their food or seize flying fish as they leap out of the water.

In active control of stability, the wings are more important than the tail, as many a bird that has lost its tail feathers has

A gliding bird alters its speed by adjusting the position of its wings. Partly folding its wings moves the centre of pressure (through which aerodynamic forces act) rearwards from A to B, behind the centre of gravity. This tips the bird's nose down and it accelerates.

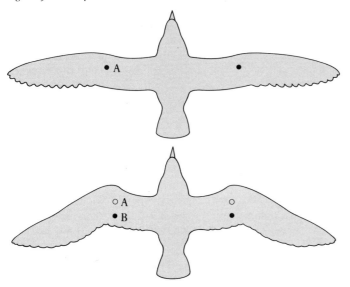

proved. If you hold a bird (perhaps a pet budgerigar or pigeon or a bird that is waiting to be ringed) and move it in the yaw, roll and pitch axes, it will make compensating movements of wings and tail that, if it were flying, would return it to a straight and level course.

For pitching control, tip the bird forward as if it was going into a dive, and the tail will tilt up so that, if airborne, its head would come up again; tip it back and the tail will drop to bring the body level.

If the wings are free, and the bird does not struggle, the same tipping movements make the wings swing forward when the body is angled downwards and backward when it is tipped up. The effect of these movements is to shift the centre of pressure (the point on the aerofoil through which lift acts). In a free glide, swinging the wings forward moves the centre of pressure forward and brings the head back up. Conversely, if the head tips up, the wings swing back to compensate. The same effect is achieved by changing the angle of attack: increasing the angle moves the centre of pressure forward.

Rolling the bird to one side makes it flap the downward wing, which in flight will result in increased lift raising it again until the wings are level. This is rather a drastic reaction and, in practice, slight flexing or twisting of the wing to increase or decrease lift on one side, as Wilbur Wright noted, are sufficient to restore equilibrium.

A free-flying bird uses its rapidly beating wings to prevent unwanted rolling but, without a tailfin, it has no natural stability against yawing movements and has to make continual corrections to prevent it from swinging like a weathercock in a gusty wind. If the bird starts to yaw, the outside wing is extended and twisted to adjust the angle of attack, thus increasing drag without changing lift, and the swing is therefore countered.

The pioneers of flight studied stability by watching large birds as they glided by and even now the best way to observe control movements is to visit cliffs where gulls and fulmars are soaring along the edge, or find an embankment where a kestrel is hovering. It is much more difficult to observe how birds manoeuvre, because they fly past quickly and their movements are so rapid, but in essence manoeuvring reverses the actions which the bird uses to gain stability.

A gliding bird's sequence of movements while turning has been described by Walter Newmark, a glider pilot quoted by James Fisher in his book *The Fulmar* (1952). He compared fulmars with high-performance sailplanes, beside which he decided gulls and kittiwakes were as clumsy as primary trainers. Newmark wrote: 'The tail muscles are extraordinarily strong and well developed, capable of warping the tail up on one side and down on the other, and at the same time twisting the whole assembly and thus putting on bank without using aileron control. The ensuing turn is then flown with the tail horizontal to the horizon and very slightly depressed.' The bank or inclination is necessary to prevent sideslip, because turning creates a centrifugal force pulling the bird outwards. Tilting the wings directs the lift inwards to compensate for centrifugal force and the tighter the turn, the steeper the bank must be.

Newmark also noted that fulmars have an alternative method of turning, in which they depress the inside wing to make a negative angle of attack. This causes the bird to bank;

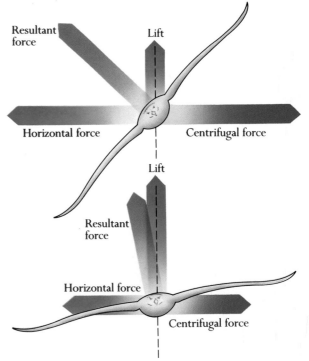

ABOVE When turning, a bird banks to prevent side-slip. It tilts its wings so that lift increases the inward force and counterbalances the centrifugal force pulling the bird sideways.

RIGHT A snowy owl using its tail as a vertical rudder, while banking.

it then twists its tail almost into the vertical and uses it as a rudder. But the main method of steering among birds in general, as revealed by photographs, is to use the oarsman's technique of pulling hard with one oar while backing with the other. Georg Rüppell filmed a redstart at 1,200 frames per second as it approached its nestbox and showed that it could spin through 180 degrees in a single wingbeat lasting 0.05 of a second. The wing on the outside of the turn beats deeply to increase its thrust while the inside wing is set at a high angle of attack to increase its drag, and the bird pirouettes. The advantage that the bird has over the oarsman is that its inner wing backs without interrupting the rhythm of the wingbeats.

A long tail assists sharp manoeuvres, especially in birds with forked tails, such as swallows, terns, bee-eaters and frigatebirds. While the widely-spread 'streamers' will enhance lift at low speed, as suggested in Chapter 2, it can be no coincidence that these birds manoeuvre with considerable ease. But none can rival the swallow-tailed kite. To quote Elliott Coues, writing in 1874 when the species was more widespread in North America: ' …the Kite courses through the air with a grace and buoyancy it would be vain to rival. By a stroke of the thin-bladed wings and a lashing of the cleft tail,

its flight is swayed to this or that side in a moment, or instantly arrested. Now it swoops with incredible swiftness, seizes without a pause, and bears its struggling captive aloft, feeding from its talons as it flies; now it mounts in airy circles till it is a speck in the blue ether and disappears.'

For most birds the tail probably plays only a minor role in steering, as is demonstrated in most spectacular fashion by the bateleur eagle. This small eagle is usually seen gliding high and fast in a straight line over the African savannahs where its 'flying wing' silhouette is unmistakeable. It has long wings, bearing more secondaries than any other bird of prey, but the tail is so short that the legs protrude beyond it. In courtship and defence of its territory, the bateleur is transformed: its name is French for tumbler and it carries out breathtaking aerobatics, with tight turns, somersaults and high-speed swerves. The tail can have little effect, but slight adjustments to the broad wings with their deeply emarginated tips throw the eagle about the sky.

Aided by its tail and feet, a black-browed albatross uses fine adjustments to the set of its wings to maintain stability as it glides slowly over the cliff in search of a place to land.

FINE CONTROL

Manoeuvres such as landing on a perch or picking up food in flight require a very fine co-ordination between the senses and muscles. Little is known about the ways birds control the fine movements of flight, but it must include the senses of balance and sight and the transmission of information from the body to the brain about the position of the wings and the airflow over them. When the first bird-like reptiles climbed and leapt in the trees, assuming the 'trees-down' hypothesis (see page 16) to be correct, they must have developed a fine sense of balance. This would then have been improved as Pro-avis took first to gliding and then to flapping flight.

Archaeopteryx fossils show that it had a well-developed cerebellum, the part of the brain controlling muscular co-ordination, and when *Archaeopteryx*' descendants lost the

A bald eagle swoops at speed, maintaining control by a fine co-ordination between the senses and the muscles of wings and tail. Its right wing is partly folded to effect a sharp turn to the right.

long stabilizing tail, the need for the brain to regulate rapid and precise movements became even more imperative. To provide the brain with the information needed to control the muscles of the wings and tail, birds have evolved excellent eyesight, while the airflow over the surface of the body is monitored by filoplumes, small hair-like feathers that are sensitive to movement.

One problem that has not been solved by pilots or birds is how to detect in advance the rising and descending air currents of thermals and other air movements (see page 101) or of turbulence. Glider pilots approaching a cumulus cloud with its associated thermal may run into a fierce down draught and be severely shaken. Neal Smith watched a flock of Swainson's hawks suddenly descend from 1,000 metres (3,280 feet) and, despite vigorous flapping, crash into trees. Turbulent wind near the ground is a cause of accidents as planes, taking off or landing, suddenly lose lift, and I have heard of a whooper swan flying steadily into a gale but falling out of the sky as it hit turbulence over a river embankment.

The ultimate in flight control is exercised by predatory birds that have to close in on a moving target at high speed and strike the target without harming themselves. Precise co-ordination is needed between eyes and wings; the bird must compute the speed and course of its victim and adjust its own trajectory to bring them together. It is like a tennis player racing across the court to return the ball, except that for the bird failure means, at best, a missed meal and, at worst, injury through an awkward collision.

The sequence of movements in an attack by a bird of prey is too fast for the human eye to register in detail but it can be revealed by slow-motion filming. In one analysis of a goshawk striking a pigeon, the hawk stopped flapping 8 to 9 metres (25 to 30 feet) from the pigeon and lowered its feet when 1.5 to 2 metres (5 to 7 feet) away. This was only about 100 milliseconds before impact, yet it still had time to fling its legs forwards so that the claws struck at 80 kilometres (50 miles) per hour, or twice the speed of the bird's body. Simultaneously, it threw its body up so that the wings and spread tail were in the maximum braking position to prevent overshooting.

To carry out such an attack, the bird has to keep its eyes on its victim throughout the attack, like the tennis player keeping his eye on the ball. The bird is aided by its ability to keep its head horizontal while its body is banking and there is evidence that birds' brains work in the same way as those of athletes. They make an overall judgement of the time to make contact with the target by detecting the changing size of the image of the approaching target on the retina of the eye. A cricketer uses this system to time catching a ball within 10 milliseconds and a bird probably works to the same level of accuracy.

Gannets catch fish by diving headlong into the sea from heights of up to 30 metres (100 feet). They keep their wings half open to help them steer as they drop, but a fraction of a second before impact they sweep their wings back and enter the water like an arrow. They must delay the sweep-back for

Gannets hunt by flying high over the sea until a fish is spotted near the surface (RIGHT), when they dive headlong into the water. Perfect timing is needed to prevent injury on impact.

A flock of budgerigars descends at the edge of a waterhole. They are well spaced as they fly in, but as the birds bunch together and drop down there is a danger of collision in the dense mass.

as long as possible to enable them to steer but not too long or the wings will be smashed by the impact. Like the bird attacking its prey in mid-air, the gannet judges the best moment for sweeping back its wings by monitoring the changing images on its retina rather than simply acting at a fixed height or time before impact. In practice this means starting the sweep-back earlier on high dives because the gannet will be travelling faster and will need more time to move its wings.

Split-second timing is also needed by birds that fly in dense flocks. A large flock taking off in a tight mass is an impressive sight. Starlings, red-winged blackbirds, queleas and other species fly up in a great mass so thick that it is almost impossible to see daylight through the beating wings and the scenery beyond is obliterated. There is little room for manoeuvre within the flock and the birds appear to be in constant danger of collision, yet they fly away, changing direction together with marvellous co-ordination.

The birds that commonly fly in dense flocks are not the most manoeuvrable. Collisions between starlings are, in fact, quite common when they leave the roost, but at this point they are still sorting out their formation. Once the flock has formed up, the birds are neatly spaced and wheel and dip without mishap. In the 1920s the statesman Lord Grey described the evening flight of starlings:

'They then fly at speed above the roosting place: a vast globe, it may be, of some thousands of birds. They fly close together, and there are many evolutions and swift turns, yet there is no collision: the impulse to each quick movement or change of direction seems to seize every bird simultaneously. It is as if for the time being each bird had ceased to be a separate entity and had become part of one sentient whole: one great body, the movement of whose parts was co-ordinated by one impulse or one will affecting them all at the same moment.'

The co-ordination of the movements is only simultaneous to the human eye. When slowed down by the cine camera, it can be seen that a ripple of movement passes through the flock, but each bird appears to react to its neighbour faster than the known reaction time of eyes, nerves and muscles would allow. The answer to such rapid reaction is, it seems, to be found in the chorus line of dancing girls. A change in step passes from girl to girl faster than the usual 200 milliseconds it takes a human to react to a visual stimulus. Each girl is watching the approaching change of step down the line, rather than taking her cue from her immediate neighbour, and so

Above Red-winged blackbirds take off in panic in all directions.

Below A moment later they have settled into a well-spaced formation heading in one direction.

times her reaction to coincide with its arrival.

In the bird flock, when one bird decides to turn, its neighbours follow suit to take up the new alignment. The birds next to them react a little later, but as the movement passes through the flock the rate of change speeds up. The birds, like dancing girls, are taking their cue and timing from individuals several places ahead of them.

The unthinking, automatic imitation of flockmates can be seen among slower birds. The American ornithologist Glover Morill Allen describes a flock of two hundred pelicans flying in formation down the Nile. At one point the leader made a small jump, for no apparent reason, as if vaulting an invisible hurdle. Thereafter, as each successive pelican reached that spot in space, it jumped in precisely the same fashion.

CHAPTER 4

STYLES OF FLIGHT

The 8,600 species of birds have taken up many styles of flying, varying the basic pattern of flapping flight described in the previous chapters. These techniques are adaptations related to the behaviour and ecology of each species. They include variations in the structure of the wings and tail to improve the mechanical efficiency, and changes in the style of flight which are usually aimed at making significant savings in the amount of energy used.

South American terns swarm above their nest. If predators approach too closely they dive to attack.

WINGS AND FLIGHT

A bird flying on a straight and level course by flapping its wings burns up energy at a prodigious rate, so it is obviously advantageous to cut down fuel consumption. Economy can be achieved by altering the shape of the wings and the style of flight. Other things being equal, flapping flight is 'cheaper' with wings of a large area, to give a low wing-loading, and a long span, to reduce induced drag, but a bird needs wings that are most suited to its lifestyle. Economy is further achieved by gliding and other flying techniques.

Long wings are only practical for birds that live and fly in open airspace, such as the albatrosses and their relatives, frigatebirds, the swifts, swallows and martins* that hunt insects on the wing, the falcons that hunt over open country and many of the seabirds. Swallows spend as much as half the day in the air, while swifts spend most of their lives aloft. Consequently there is a need for effortless flight and this is served by long wings and low wing-loading (the swallow's

---

*Swallows and martins are close relatives in the family Hirundinidae and are often lumped together as 'hirundines' by ornithologists. The swifts are a very different group of birds which are probably most closely related to the hummingbirds.

wing-loading is half that of the stub-winged wren).

There are several reasons for abandoning the long, narrow, aerodynamically efficient wing. Large birds are limited in their wingspan because a long wing would be structurally weak and difficult to control, so some large gliding birds, such as vultures, cranes, storks and pelicans, have opted for shorter wings with emarginated tips to reduce drag. As Colin Pennycuik has pointed out, an albatross-shaped vulture would be a better glider, but Nature is full of compromises and a large bird with long, narrow wings would find take-off extremely difficult, whereas wings of low aspect ratio with wingtip slots develop high power at low speed. Other birds need to fly fast and use high aspect ratio to reduce induced drag, but an altogether smaller wing gives a higher wing-loading. This describes the wings of fast-flying ducks and waders, which are, however, poor gliders.

Right The slender, pointed wings of the bee-eater are the hallmark of a bird that forages on the wing in open airspace.

Below The great pied hornbill's broad, round wings are suited for manoeuvring through the forest canopy.

	Wing loading (gm/cm^2)	Aspect ratio	Stalling speed (km per hour)
Pied flycatcher	0.13	5.9	13.7
Wren	0.24	6.9	18.4
Kestrel	0.35	7.7	22.3
Frigatebird	0.50	12.6	26.6
Pigeon	0.52	6.3	27.3
Griffon vulture	0.70	6.2	31.7
Pheasant	1.04	4.6	38.5
Mallard	1.20	9.0	41.4
Wandering albatross	1.37	18.7	44.3

This table shows the relationship between wing-loading, aspect ratio and minimum flying speed. Stalling speed is that at which flapping is needed to keep the bird from falling out of the air.

Note the high wing-loading and fast flight of the mallard, and the high aspect ratio, low wing-loading and slow speed of the frigatebird. The wandering albatross is a fast glider, with a high wing-loading and aspect ratio, compared with the griffon vulture which is slower but turns in a tighter circle. The pied flycatcher is about the same weight as a wren but its low wing-loading makes it more manoeuvrable. The pheasant's high wing-loading and low aspect ratio, combined with its large flight muscles, are an adaptation for life in the woods.

Birds living among trees would find long wings a hindrance and they have traded the advantage of a long wing for the convenience of a short wing. The wings of woodland birds are typically blunt, as in woodpeckers, jays, magpies, hornbills, toucans and wrens, and it is interesting to note the contrast between other woodland species and their open-country relatives. For example, the starling is a bird of pastures and lawns and it is not skilled at aerobatics. Its silhouette overhead is a characteristic 'arrowhead' with pointed wingtips and contrasts with the rounded outline of its relatives, the glossy starling and shining starling, which have to manoeuvre through the forests of Polynesian islands. The same pattern holds for the short-eared owl that hunts over open country and the tawny owl, which is its woodland counterpart. The latter has short, broad and rounded wings with a higher wing-loading than the short-eared owl and it's rapid, shallow wingbeats are an aid to weaving through confined spaces among the trees.

Flightstyle keeps swallows, martins and swifts out of the trees where their place as hunters of flying insects is taken by the flycatchers. Similarly, the falcons and harriers of open country are replaced by goshawks and sparrowhawks. Although the wings of woodland and forest birds are shorter, and also broader to keep the wing-loading down and increase manoeuvrability, the resulting low aspect ratio increases the cost of flight. The woodland flycatchers, however, keep the cost of flying down by waiting on a perch and making short sallies to seize passing insects. This trend is taken even further

RIGHT A dipper flies with rapidly, whirring wings, proceeding up and down the river in short spurts.

WING SHAPE AND FLIGHT STYLE

The long, pointed wings of albatrosses and petrels and the broad, fingered wings of harriers and vultures are efficient for soaring in open space. Comparatively small wings and fast flight are best for the long migrations of the plovers and other waders. In contrast, the short wings of jays and catbirds are used for manoeuvring through trees. The pointed wings of the migratory sedge warbler are more suited for long-distance flight than the rounded wings of the moustached warbler.

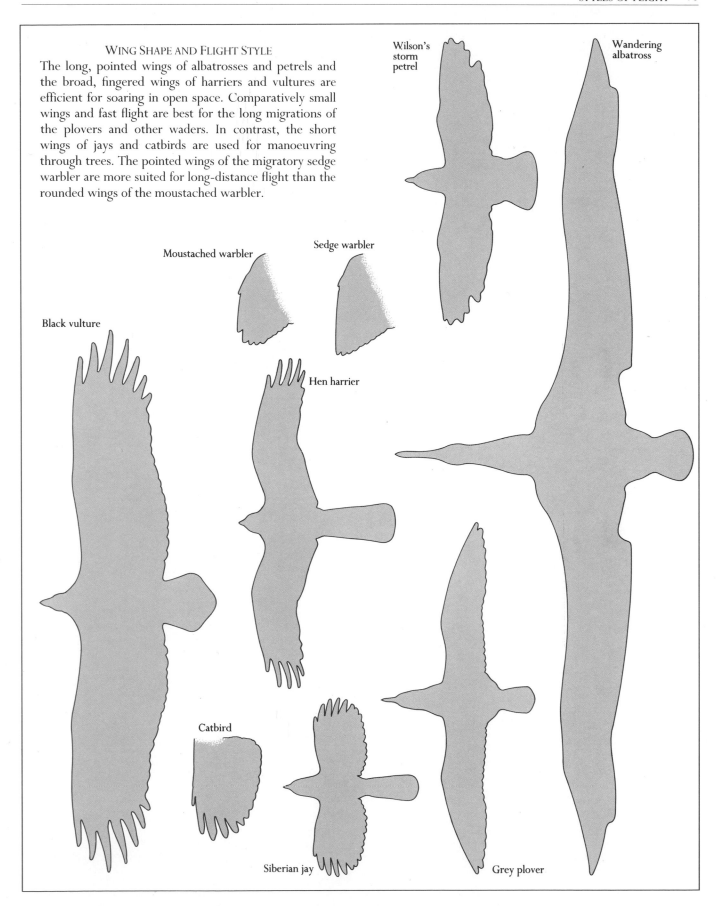

Wilson's storm petrel

Wandering albatross

Moustached warbler

Sedge warbler

Black vulture

Hen harrier

Catbird

Siberian jay

Grey plover

by the tits which feed amongst foliage. Their aspect ratios are even lower and the cost of their flight is higher, but they spend most of their time clinging to leaves or hanging from twigs.

The high-aspect-ratio wings of swallows and swifts have a second advantage: they are also the best design for long-distance migration flights. The same dual role is seen in the long-winged long-eared owl which is a migrant, while its relative, the round-winged tawny owl, keeps to its territory throughout its life.

There often has to be a compromise in wing design where birds have conflicting needs. Many warblers, for instance, feed on insects by flitting among the foliage in slow, almost hovering flights, but twice a year they fly thousands of kilometres on migration. Ideally, they need two sets of wings: one for slow flight while feeding and one for faster, economic travelling. The goldcrest is a warbler that plucks tiny insects from leaves and sometimes migrates long distances, including hops across the sea. Its low-wing loading makes it suitable for slow flight and hovering among dense conifer foliage, but the wingtips are pointed as a concession to the need for economy on sustained migratory flight.

LEFT A tawny owl takes off from the woodland floor. Broad, rounded wings of low aspect ratio are used for flying in the confined airspace of its territory.

ABOVE Bar-tailed godwits, like most waders, are long-distance specialists that migrate to breeding grounds thousands of kilometres away from their winter habitat. Slender wings contribute to the economies needed for crossing expanses of ocean.

BELOW The short wings of the guillemot are used for 'flying' under water, but they impose constraints on flying in the air.

FREE-WHEELING

We have seen how birds can adjust their speed to fly with the minimum requirements of power (Chapter 3), but there are several tactics for reducing energy expenditure further. One is simply to stop flapping and glide. This makes a saving of about 10 or 20-fold because during gliding energy is only required for holding the wings outstretched and making the small movements needed for steering and stability. All birds glide at times, either when they are losing height before landing or if they are heading into a wind strong enough to give them flying speed, but many birds are good gliders and regularly intersperse bouts of gliding with flapping flight even in still air.

Flapping and gliding is called undulating flight. The duration of the gliding phase depends on wind speed and direction, and the bird flaps in bursts to regain the height lost

BELOW White pelicans demonstrate the technique of undulating flight, in which they flap and glide in energy-saving locomotion. The length of time that a bird can continue to glide depends on the weather conditions at the time.

ABOVE When not in a hurry, the herring gull flies economically by alternately flapping and gliding. This is more efficient than continuous slow, flapping flight, which does not use the muscles effectively.

during the glide. Undulating flight is commonly practised by fairly small birds such as starlings, bee-eaters, swifts and swallows, medium-sized fulmars, gulls and crows, and large birds such as herons, cranes and eagles.

Some of the finest exponents of undulating flight are birds of prey and owls that hunt over open ground. Their strategy is to stay aloft as long as possible while they scrutinize the ground for the slight movements that give away the position of their prey. Speed is not important, rather the reverse because moving slowly gives them more time to make a thorough scan. Harriers and short-eared owls that use this hunting technique have low minimum power speeds, compared with similar-sized jackdaws, and require less energy for flapping flight. Undulating flight gives them increased savings of 10 to 20 per cent.

BOUNDING FLIGHT

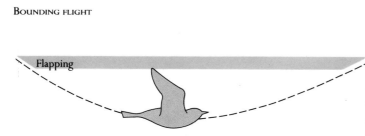

Flapping

Wings closed

(*Not to scale*)

Small birds probably save energy by cruising with a bounding flight, alternately flapping and closing their wings. The aerodynamics of their small size makes this pattern of flight more economical than flapping and gliding.

BOUNDING

Small birds do not use undulating flight, and never glide for any appreciable distance. Instead they use a bounding flight in which bursts of flapping alternate with short phases of hurtling through the air with wings closed. The bird flies upwards, then folds its wings and curves over in a ballistic trajectory, flaps again to accelerate and flies upwards once more. Bounding birds include warblers, tits, finches and sparrows, and many others up to the size of a small owl or parrot. Bounding can be seen in flocks of finches crossing open ground and is very exaggerated in wagtails and woodpeckers, but it is never used when the birds are disturbed and escaping as fast as possible. The best time to see bounding flight is on a windy day when small birds are fighting to make headway. Their progress is so slow that they look as if they are bouncing on the end of a spring.

The theory of bounding flight, as yet unproven, is that it saves energy. Small birds typically have short, broad wings of low aspect ratio, which means that induced and profile drag is high. If these birds attempt to glide while travelling fast, they will save the induced power of flapping but continue to incur a high cost of profile power from their outspread, relatively large wings. So, where larger birds 'freewheel' by gliding in undulating flight, small birds fold their wings to cut down drag in bounding flight. A second advantage is that, for anatomical reasons, small birds cannot vary the power output of their flight muscles to any great extent. The muscles are most efficient when contracting at a particular rate, which may be that used when taking off and accelerating. So the best way to reduce energy consumption in normal flight is by bursts of flapping at the optimum rate for muscle efficiency, in which the bird accelerates, combined with 'freewheeling' with the wings folded, to obtain a lower average power output. The economy of travelling in bursts of power was demonstrated by an American test for motor cars. The car that went farthest on a gallon of petrol was a high-powered machine that accelerated to top speed at maximum power output, then coasted with the engine off until it was down to a crawl before racing off again.

The green woodpecker is one of the largest birds to practise bounding flight. Its trajectory rises and falls as bouts of flapping alternate with wing closure.

SOARING

Great savings in the cost of flight can be made if the bird can extract energy from the atmosphere, in other words, if it makes use of air currents either to help it along or to provide lift and counteract gravity. A tailwind is obviously a great help during long migration flights and its effects are discussed in Chapter 5, but airflow can provide extra lift to offset the pull of gravity and reduce the dependence on flapping flight. When this happens the bird is soaring.

A gliding bird is using the force of gravity to provide motive power and consequently loses height. It has to flap to regain height, or eventually it will hit the ground, but its gliding range can be extended by extracting energy from moving air and using it as 'fuel'. If the bird can extract enough energy from the air it can remain airborne for long periods.

TOPOGRAPHIC SOARING

Wherever the wind meets an obstacle it is thrown into eddies of turbulent air, spinning in all directions but often with enough of an upward current for birds to use. A ship's forward movement through still air is sufficient to set up eddies above the stern in which gulls and albatrosses can hang and be drawn along while waiting for food to be tossed overboard.

Buildings, woodland edges and hedges also provide a source of eddies, but there is more opportunity along the windward sides of hills, sea cliffs and islands where the wind is displaced upwards in a steady current, as well as being thrown into eddies. In hill country this 'slope soaring' is important for birds of prey, choughs, ravens and mountain specialists such as the snow pigeon of the Himalayas, or any bird with a low

ABOVE The eddies flowing around the hull and superstructure of a ship provide updraughts in which herring gulls soar effortlessly.

RIGHT Gannets glide over the Bass Rock colony, off the Scottish coast, with the aid of wind blowing up the cliff face.

wing-loading. Gliding along the line of the slope enables them to travel across large areas in search of food or to cover distances easily on migration.

Some birds that are hopeless gliders in still air can make use of air currents. Strong winds off the sea create such powerful updraughts along cliff faces that even heavy cormorants and small-winged auks, such as guillemots and puffins, can join fulmars and gulls soaring around their colonies in stormy weather. Downwind of isolated hills and islands, the airflow may be thrown into standing waves, called lee waves, which provide extra regions for soaring. Seabirds, for instance, can work their way upwind to island colonies by ascending in the lee waves and gliding from one to another.

Eleanora's falcon uses hovering as a hunting technique during the latter part of the breeding season when it changes from hunting insects to pursuing birds. It is unusual because it breeds late in the year and its nestlings are fed on the migrant birds that swarm across the Mediterranean to spend the winter in Africa. The migrants travel by night and reach their destination at daybreak. If they have arrived near a colony of Eleanora's falcons they must run the gauntlet of parent birds waiting to feed their offspring. The falcons gather in the airspace around the colony and, making use of the stiff

Wind is thrown into waves as it passes over hills, sea cliffs and even buildings, hedges and embankments. Birds make use of the updraught for slope-soaring. They glide along the faces of hills and cliffs to cover long distances and search for food. Sometimes the lee waves can be used to travel farther afield by gaining height between glides. Swifts and other birds fly in rotating eddies to catch insects which have gathered there in swarms.

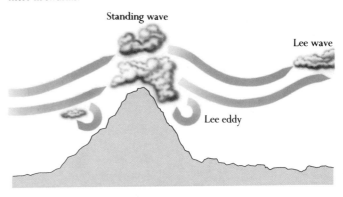

Standing wave

Lee wave

Lee eddy

BELOW A giant petrel soars on the strong winds that race endlessly over Antarctic waters. Note the head held horizontally while the body banks at right angles.

updraught off the cliffs, they hang in the same position for up to 10 to 15 minutes or glide to and fro along the cliff face. From this vantage point the falcons spot approaching migrants and swoop down on them. Hartmut Walter, who has studied Eleanora's falcon on the island of Praximada, off Crete, describes the 'Praximada falcon wall', the population of Eleanora's falcons strung along the coast from sea level to 1,000 metres (3,280 feet), several kilometres wide and deep, in a unique live bird-trap.

A rather different form of slope soaring is important for oceanic birds. When a wind is blowing, it outpaces the waves it is throwing up and forms an upward flow of air over them. Seabirds use this to soar across the wind by gliding along the line of the waves and progress upwind or downwind by shifting from wave to wave.

Albatrosses continue to glide close to the surface when the wind has dropped to a flat calm, a puzzle that intrigued William Froude when he sailed to Cape Town in 1879. Froude is best known for his invention of the tank in which

The secret of soaring is control of flight. This fulmar is using its wings, tail and feet to hang almost motionlessly on a breeze that is coming up the cliff face.

scale-model ships simulate the behaviour of the full-size hulls, but he applied his engineer's mind to the albatrosses that flew around his ship without flapping:

'We were sailing ... with a well-marked swell of about 8–9 second period, and varying from 3 or 4 feet to 8 or 9 feet from hollow to crest. The speed of such waves would be from 24 to 27 knots. Under these conditions the birds seemed to soar almost *ad libitum* both in direction and speed .... Now skimming along close to the water, with the tip of one or other wing almost touching the surface for long distances, indeed now and then actually touching it.... The action was the more remarkable owing to the lightness of the wind, which sometimes barely moved our sails.'

Froude realized that long after a storm has abated and the wind has died away, a swell continues to run and displaces the air ahead of it. A simple calculation showed him that air is pushed up the leading side of the swell at about 1 metre per second, which is enough to counteract the sinking rate of a gliding albatross and keep it gliding when there is no wind.

THERMAL SOARING

At the turn of the century E.H.Hankin studied the soaring vultures and kites that gathered to scavenge at the abattoir

outside Agra in India. He invented an ingenious method for recording the birds' flight patterns. A mirror was set up so that he could observe the image of a bird in it and mark its position with a pen at intervals (given by the beat of a metronome). The series of ink blobs were then preserved by being blotted onto a sheet of paper.

Hankin noted that white-backed vultures took off by flapping and glided only after they had gained height. In the early morning they started to glide at 10 to 20 metres (33 to 66 feet). Later, in the heat of the day, they could glide when they were only 4 to 5 metres (13 to 16 feet) off the ground. The heavier vulture species took off from their overnight roosts later in the morning. Hankin reasoned that there was some property of warm air that made it easier for them to start soaring, but he could only say that the conditions resulted in the air being 'soarable'. It is now well known, however, that vultures and many other birds – as well as glider pilots – make use of rising currents of air called thermals. The Californian condors that had difficulty with the headwind (see page 47) used a thermal to climb out of trouble 'as a balloon might rise', becoming lost to view in the clouds and emerging again above them.

When the ground is warmed by the sun, the cool air in contact with it heats up. Warm air is lighter and rises through the atmosphere as a bubble of air or thermal, which can be visualized as a ring-shaped vortex spinning on its axis. A draught of air is drawn up through the centre of the vortex, travelling faster near the centre, and spreads out at the top. A thermal is most likely to form where there is a surface that will heat more quickly than the surrounding ground. This could be a rock outcrop, dry sand, a building or a clearing among trees. As the thermal rises into cooler air its water vapour may condense to form a cumulus cloud, hence the layer of scattered 'fine weather cumulus' that builds up on warm days. If there is a breeze, a succession of thermals are generated from a single source and drift downwind, each with its own cloud, to make a 'thermal street'. The thermals

eventually reach a ceiling and stop rising when they have lost their lift. As the sun's heat diminishes in the afternoon and the ground cools, the production of thermals weakens and they disappear in the evening.

It is sometimes said that thermals do not form over the sea but, while this is generally true, they occur in the tropics where the trade winds bring in cool air over warm water, while in winter they form at high latitudes where cool air from the frozen land flows over relatively warm sea. Whereas land cools at night, the sea remains warm and the thermals do not die down in the evening.

An unusual form of thermal generation has been witnessed where storks migrating across the Sahara have soared above flares of burning gas at oil wells, and Norman Elkins has written in *Birds and Weather* (1983) that he believes that gulls may start their own thermals. If there is a marked temperature difference between cool air and warm sea, and no wind, a large flock of circling gulls can set the air rotating and rising.

The object of the soaring bird (and the glider pilot) is to find a thermal and circle in the centre. As it enters a thermal, a bird that has been gliding with a reduced wingspan and closed wingtip slots, slows down, spreads its wings, opens its slots and fans its tail. It is now set for its minimum sinking speed, i.e. the speed at which it loses height most slowly.

A soaring bird's behaviour depends on the strength of the thermal. It can turn in a tight circle in the centre of a strong thermal where fast-rising air compensates for the reduced lift on the steeply banked wings, but where the air is rising more slowly it describes wide circles and gets more lift from horizontal wings. Either way, it floats, seemingly cut loose from gravity. Alternatively, the bird can climb or sink in the thermal by adjusting its circling so that its sinking speed is lower or higher than the updraft. Once aloft and circling in its envelope of warm air, the vulture or any other soaring bird maintains height with a minimum energy expenditure or it can leave the thermal and travel cross-country by peeling off and gliding to the next.

As the ground warms up in the sun, bubbles of warm air rise as thermals and drift downwind. Their position is marked by small cumulus clouds and by birds circling in them to make use of the rising air.

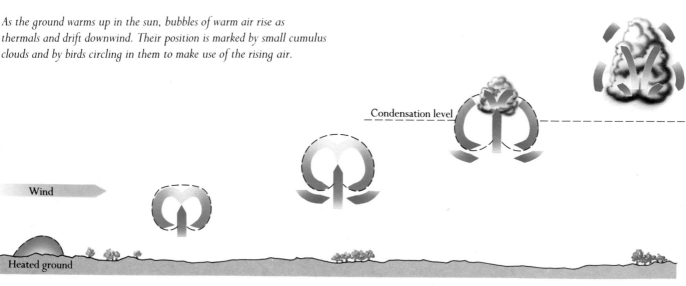

Vultures and marabou storks circle together in a thermal. These birds rely on soaring to give them an easy way of patrolling the countryside in search of carrion.

Large, broad-winged birds such as the bigger vultures, condors, pelicans and storks may be dependent on thermals for flight activity for much of the time, and the use of thermals explains Hankin's observations of the vultures at Agra. The heavier vultures with higher wing-loadings were the last to take off in the morning because it takes time for the ground to warm up and develop thermals strong enough to support their weight. Also, when the thermals have intensified, they need to use less flapping flight before starting to soar. Colin Pennycuik has compared the soaring abilities of the brown pelican and the black vulture. Their wing-loadings are similar but the pelican climbs in a thermal more rapidly because of its higher aspect ratio (9.8 compared with 5.8).

Marine thermals in tropical seas are the province of frigatebirds. They are not dependable enough for land birds to rely on when making sea crossings, but frigatebirds have extremely low wing-loadings and very buoyant flight, so they can remain aloft in weak but persistent marine thermals and forage far from land.

Another form of lift generated by atmospheric conditions is the result of sea breezes and thunderstorms. A sea breeze occurs when the land warms up during the day, causing the air over it to rise and suck in cooler air from the sea. A good time to see this is on a day of glassy calm; the edge of the sea breeze is marked by a dark trace on the water and gulls soaring along its length, and as it moves inland a line of cumulus clouds may develop. A similar sort of lift forms around the outskirts of a thunderstorm where cool air descends through the cumulonimbus thunder cloud and spreads out, pushing up the warm air around it. Swifts

BELOW Sea breezes blow because the land warms up faster than the sea during the day. Warm air rises over the land and flows out to sea at high level to be replaced by cool sea air. Birds soar in the line of rising air at the boundary.

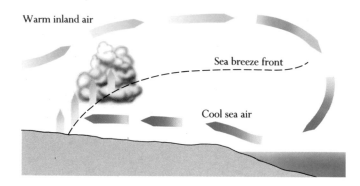
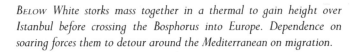

Warm inland air

Sea breeze front

Cool sea air

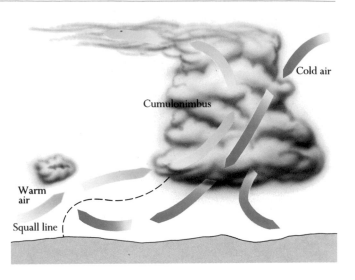

Cold air

Cumulonimbus

Warm air

Squall line

BELOW White storks mass together in a thermal to gain height over Istanbul before crossing the Bosphorus into Europe. Dependence on soaring forces them to detour around the Mediterranean on migration.

ABOVE Thunder clouds sometimes produce a strong downdraught. Where this meets the ground and spreads out, it displaces the warmer air ahead of it. The air that is forced up is used by birds for soaring.

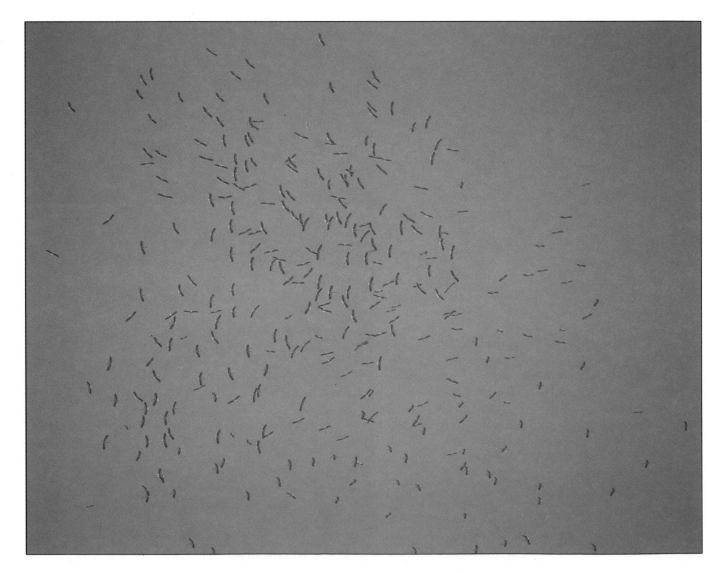

regularly use these conditions to feed on the insects and tiny spiders swept up in the rising air, but glider pilots have reported meeting kestrels, buzzards, gulls and herons using the same rising air over southern England.

DYNAMIC SOARING

Albatrosses have always excited the admiration of seafarers for their seemingly endless flight around the stormy latitudes of the southern ocean, but the way they ride the wind without flapping was a puzzle. Not surprisingly, some fantastic ideas were spawned and I like the one repeated by the eighteenth-century poet Oliver Goldsmith that 'at night, when they are pressed by slumber, they rise into the clouds as high as they can; there, putting their head under one wing, they beat the air with the other, and seem to take their ease'. However, Goldsmith does go on to describe how the albatrosses 'alternately ascend and descend at their ease'. This is a succinct description of the form of flight known as dynamic soaring. Whether leisurely in a gentle breeze or speedily before a gale, the albatrosses drop to the waves, then turn and climb again to their original height. The pattern is repeated for hours on end, wherein lies the fascination of watching them. The phenomenon can be explained by thinking of a car that freewheels down a hill. The impetus will carry it part of the way up the other side of the valley but drag slows it down and it will eventually come to a halt below the top of the hill. The gliding albatross overcomes drag and rises to its original height, so it has found some extra energy to carry it the last few metres.

The first suggestion for a physical explanation of this apparent feat of levitation came from Lord Rayleigh, who proposed in 1883 that the albatross extracts energy from the wind gradient that exists over the surface of the sea. Friction slows the wind when it is in contact with any surface, so windspeed increases with height up to around 15 metres (50 feet). Lord Rayleigh's theory was that the albatross glides

A black-browed albatross climbs into the wind without wingbeats. Using the momentum of a shallow dive to propel it upwards, it gets extra energy from the increasing windspeed.

Above A wandering albatross, with a black-browed albatross behind it, soars over the steep swell, propelled by the almost constant winds of the southern ocean.

Below Albatrosses and other birds use the wind gradient above the sea for soaring. By zig-zagging across the wind, they fly long distances with the minimum effort.

Increasing wind speed

Travellers in the southern ocean have always known that the wandering albatross deserves its name. They have seen it following ships many hundreds of kilometres from its nesting colonies on Antarctic islands and evidence showed that some albatrosses were travelling up to 1,800 kilometres (1,120 miles) from the nest in search of food. This is only possible because the almost continual winds at these latitudes enable wandering albatrosses to soar with hardly a wingbeat over vast stretches of ocean.

French ornithologists have now shown how amazing the flight capabilities of the wandering albatross are by fitting six of the birds with miniature radio transmitters which can be tracked by an orbiting satellite. They found that, between relieving their mates on the nest, these birds may travel up to 15,000 kilometers (9,320 miles) and cover the ocean at speeds of up to 80 kilometres (50 miles) per hour.

On leaving their nests on the Crozet Islands, south-east of South Africa, the albatrosses let the prevailing Westerlies sweep them downwind. They travelled mainly by day, probably because they were feeding at night when squid come to the surface.

The study solved the longstanding puzzle of how albatrosses get back to their nests against the prevailing wind. They either tack back, zig-zagging across the wind, or they fly in a large loop and make the final approach downwind again.

However, when the wind drops, the albatrosses are becalmed and sometimes do not move significantly for several days. It does not matter if an albatross is held up by contrary winds during the incubation period, because its mate can survive on the nest for a long time. Strategy changes after the egg has hatched, when shorter flights of two to four days are made so that the nestling receives frequent meals.

downwind, gaining groundspeed, then turns sharply and heads up into the wind. Its groundspeed now falls off but, because it meets an increasing wind velocity as it climbs, its airspeed does not slow down so fast. In this way it regains its original height before its speed drops below its minimum gliding speed, then it can turn and repeat the manoeuvre.

Dynamic soaring is often assumed to be the only, or at least the most important, means of soaring used by albatrosses, but Lord Rayleigh did not believe this. He wrote: 'A priori, I should not have supposed the variation of velocity with a height to be adequate for the purpose; but if the facts are correct, some explanation is badly wanted.' The explanation is that the albatross is assisted by slope soaring along the waves, aided by ground effect (see page 108), which gives it the extra impetus for the climbing phase. The situation is complicated because the theory of dynamic soaring assumes that albatrosses fly downwind and along the waves. However, from the stern of a ship birds can be seen flying in all directions.

Dynamic soaring is not confined to albatrosses. A similar pattern of flight is used by the smaller species of petrels and shearwaters, as well as gulls and gannets, although they more frequently have to resort to flapping to regain height. Moreover, although dynamic soaring is always described as taking place over the sea, a wind gradient occurs on land so the technique should be available to land birds, especially over open, level ground. I have noticed a variety of birds – terns, skuas, gulls, waders, herons, and swallows migrating over the Sahara – flying with a zigzag, undulating passage that recalls the dynamic soaring flight of albatrosses. They do not glide but they could be using the wind gradient to save effort.

GROUND EFFECT

The skimmers are unique birds. The three species live alongside tropical lagoons, lakes and rivers, and are named after the way they fish by flying to and fro with their beaks ploughing a furrow through the surface of the water. The lower part of the bill, the mandible, is one third to one quarter longer than the upper part. As soon as the mandible touches a fish or crustacean, the bird's head drops and the bill snaps shut to trap the prey and lift it clear, with barely a check in the skimmer's flight.

The skimmer's neck is cushioned against the shock of ramming into its prey, and its 'ploughing' flight is assisted by the knife-edge shape of the mandible set with small ridges which are believed to reduce drag. Nevertheless, it is surprising that a skimmer gives such an impression of effortless flight, with long glides between short bouts of shallow wingbeats.

Skimmers spend a considerable part of each day flying over the water in search of food and they reduce the effort of flight through the use of ground effect. This is the phenomenon underlying the hovercraft, or ground effect vehicle, but it can also be used by aircraft flying at very low altitudes; the airflow

The black skimmer is assisted in its peculiar method of feeding by the ground effect which reduces the effort of flying.

A black-browed albatross hugs the surface of a glassy sea. It is probably using ground effect to extend its glide.

funnelled between wing and surface reduces induced drag and the pilot can throttle back to conserve fuel. Ground effect comes into play when the wings are less than their span's distance above the surface and it increases as the gap narrows. In World War II long-distance marine reconnaissance planes were able to extend their range by flying just above the waves.

Ground effect works equally well for birds flying close to the ground or water surface. With less drag, they use less energy and either flap less powerfully to maintain speed or glide farther. Because the effect is proportional to wingspan, it is more likely to be of use to birds with long wings. Skimmers have a span of about 75 centimetres (2½ feet) and, by gliding about 1 to 8 centimetres (½ to 3½ inches) above the water, they save substantially on the power needed for level flight because they will glide at a shallower angle and need flap less frequently to restore height. This makes a significant saving in the amount of food a skimmer has to find and will cut down the time it has to spend foraging. The drawback is that the skimmers' peculiar method of feeding only works in calm water and their distribution is therefore restricted to tranquil places such as sheltered lagoons and lakes.

It seems likely that other birds benefit from ground effect,

when either foraging or commuting between nest and feeding ground. Several kinds of seabirds, such as pelicans and cormorants, albatrosses, petrels and their relatives, regularly fly at a very low level. The shearwaters get their name from appearing to shear the water with their wingtips as they glide only a few centimetres above the water. Calculations made for brown pelicans which have wingspans of 210 centimetres (7 foot) and glide about 30 centimetres (1 foot) above calm water, suggest that they halve the induced drag on their wings. At intervals they climb to about 50 centimetres (20 inches) so they can flap their wings and presumably continue to gain a reduced benefit from ground effect.

Ground effect must also be of value to large birds that need to run and flap to reach flying speed at take-off. Albatrosses, vultures, auks, divers and swans spring to mind as birds that labour up into calm air. They can get airborne before they have reached full flying speed and they keep low, maintaining the ground effect and gaining airspeed so that, when they climb out of the influence of the ground effect, their forward passage will provide enough lift to keep them up. Without enough lift they would crash, as has happened to many an overladen aircraft at take-off. Ground effect will also allow these birds to maintain enough lift to keep them in the air at lower speeds, so they can slow down on the approach to a landing and touch down with less of a thump.

ABOVE Snow geese flying in echelon formation over New Mexico. By maintaining formation, small savings in the cost of flight accrue during the long migration from the Arctic regions.

LEFT Blue-eyed shags fly in classic V-formation. These heavy-bodied, inefficient fliers proceed in formation when travelling over the sea in search of shoals of fish.

FORMATION FLYING

In *The Fowles of Heaven*, written about 350 years ago, Edward Topsell described flocks of cranes flying 'triangularlie like the Greeke Lambda [Λ] pointed before and forked behinde'. This is a better description than V-formation, which points in the wrong direction. Topsell went on to remark that the 'Captayne or foremost' crane changes places at intervals because it was believed that the cranes rested during the flight by laying their necks on the backs of the birds in front. Without this advantage the leader soon tired and had to be relieved.

This notion is not as silly as it might seem. Aerodynamic theory suggests that birds flying in V or echelon formation could be saving energy by flying in each other's wake. As well as producing a 'downwash', a wing creates an 'upwash' that spreads laterally from near the tip. It is greatest behind the wing and just beyond the tip, so a bird flying in its neighbour's upwash will get a reduction in drag and reduce the power needed to fly. There is even upwash ahead of the wing, so the leader will get some benefit.

There is an authentic story from the Korean War of an aircraft containing an unconscious pilot being escorted back to

White pelicans glide over the Danube delta. The energy saving from formation flying is increased if the wingtips overlap those of neighbouring birds.

base by two others with their wingtips overlapping those of the distressed aircraft. It has often been said, however, that geese, gulls, pelicans and cranes do not align themselves in formation with sufficient proximity and accuracy to take advantage of their neighbours' upwash in this way. Calculations now suggest that this does not matter; although the greatest support from a neighbour is obtained from overlapping wingtips and the lifting effect decreases with increasing separation, there is sufficient upwash to be significant even in a rather ragged formation. For migrating birds, small savings would accrue into a worthwhile benefit over a long distance.

The angle of the V does not affect the advantage to the birds but cranes characteristically fly with their heads opposite their neighbour's wingtip, apparently for better vision. Power savings are also greatest if the birds are flying in the same horizontal plane, but cranes are staggered vertically, each bird flying slightly higher than the one in front. Presumably the slight loss in aerodynamic benefit is offset by a better field of vision for the birds.

The upwash effect also depends on the size of the flock. A bird in a group of five benefits more than one in a group of three, but the increment in benefit progressively decreases as the flock expands. It is a disadvantage to be the leader but worse to be at the edge of the formation, so the 'Captayne' crane is not the only bird to profit from changing positions.

HOVERING

Hovering is a very expensive form of flying. From the power-speed curves described in Chapter 2, it is clear that a bird with little or no forward speed has to support itself by flapping hard and directing a downward flow of air sufficient to balance its weight. Most small birds can hover for a short time, with much deeper and faster wingbeats than normal, but their muscles cannot generate enough power on a continuous basis.

Some birds hover for a few seconds to pick up food that is awkward to approach, such as berries hanging from stalks, insects sitting on leaves, or titbits on a bird table. House sparrows, starlings, blackbirds and others will also hover to peck at hanging fat, although they are not well designed for this kind of flight and the effort must be great.

A few birds can switch from hovering to flying backwards. When Georg Rüppell was filming his redstart (see page 82) he found that it not only hovered much better than other small birds, but could also fly backwards for a short distance. From a hovering position, the inner wings were tilted so far back that their lift drew the bird backwards, while the outer wings were twisted to maintain an angle of attack that continued to provide upward lift.

Larger birds use hovering to give them a platform for scanning the ground below. Buzzards, ospreys, short-eared owls, snowy owls, terns, skuas and kingfishers use this form of hunting, but the habit has become a speciality for the kestrels. It has made the kestrel the most easily recognized bird of prey and gave rise to names once current in southern England: windhover, hover hawk and wind fanner.

The typical image of a hovering kestrel is of the bird facing into the wind, alternately flapping and hanging on fixed wings. It puzzled W.H. Hudson when he watched a kestrel hovering in a strong wind. How was it that 'in the short intervals, when the outspread wings became fixed and motionless, the bird was not instantly blown ffroomm iits posiition?'

Neither the kestrel nor other large birds that hang in the air are hovering in the strict sense; they are flying or gliding slowly into the wind so that their airspeed matches the windspeed and their groundspeed is zero. This is often called wind-assisted hovering or, simply, wind hovering. The amount of flapping involved in wind hovering depends on the airspeed. In still air a kestrel flaps hard with deep wingbeats and it hangs almost vertically, its long tail spread for extra lift. With increasing windspeed, its body becomes more horizontal and its wingbeats diminish until it can poise with only small movements to adjust to changes in the airflow.

These birds use slopes and cliffs to provide updraughts, which is why kestrels are frequently seen hovering over

Heading into the wind, a kestrel hovers with winnowing wings and tail spread while it scans the ground for prey. Hovering, although strenuous, allows it to hunt over open ground.

Above An Arctic tern hovers over the photographer, alarmed at the disturbance near its nest.

Left A Wilson's storm petrel 'walking on water' in a unique form of hovering. Dipping the feet into the water improves its gliding ability over calm water.

motorway embankments, looking for voles. In a breeze, a kestrel hovers for a few seconds to scan the ground below; if nothing is spotted, it swings downwind and glides steeply, gaining enough momentum to turn back into the wind and climb again to regain its original height, rather like a soaring albatross (see page 105). If there is a strong wind it can search a strip of ground, moving slowly upwind, with scarcely a wingbeat.

Hovering is employed by several birds to facilitate hunting over open ground. On the Arctic tundra, where knolls and cliffs provide limited vantage points, long-tailed skuas, Arctic terns and snowy owls mix their slow, low-level flight with the same hovering and dipping pattern as the kestrel, as do pied kingfishers, ospreys and short-toed eagles in warmer countries. Although hovering with flapping is expensive – the American black-shouldered kite, which hunts like a kestrel, hovers for one third of its hunting time (about six hours a day), using up half its daily energy budget in the process – it gives birds the opportunity to extend their search for prey into areas where they cannot watch from a perch, so the extra effort is repaid by success in their hunting.

STORM PETRELS

The storm petrels probably gave the large petrel group of seabirds their common name through an allusion to St Peter walking on the water. They feed on small animals which they pluck from the water while pattering and sometimes standing on the surface on outstretched legs.

This 'walking on the water' is unique to the storm petrel family. It is a form of hovering, or flying slowly into the wind, in which the feet are trailed in the water like sea anchors. When normal slope soaring is impossible because the sea is calm and there are no waves to create updraughts, storm petrels would be able to glide for only short intervals between flapping (although the ground effect helps). However, if they dip their long legs in the sea, their webbed feet drag through the water so that they are behaving like a kite hovering at the end of a string.

When looking for food, a storm petrel searches large areas of the sea surface economically by letting the wind blow it along, then inspects and poises over its prey by pattering fast enough to remain stationary prior to dropping down to seize it. This mixture of pattering, flapping, soaring over the tops of the waves and landing on the surface gives an impression of restless searching and pecking that recalls their other old sailors' name of Mother Carey's chickens.

HUMMINGBIRDS

Hummingbirds are the only birds that hover for a prolonged period in still air. In an experimental situation, a hummingbird hovered continuously for 50 minutes. Normally, they hover in short bouts and the effort is amply rewarded by the rich nectar which they sip in the process. The wingbeat frequencies needed for hovering range from 10 per second in the starling-sized Andean giant hummingbird to an incredible 80 per second in the amethyst woodstar. In this unique style of flapping flight they are behaving rather like helicopters except that, Nature having failed to invent the wheel, the hummingbirds' wings beat to and fro while the helicopters' rotors revolve.

A hummingbird wing is a rigid structure in which the arm bones are reduced so that the wing is almost all hand, and the elbow and wrist joints are locked inflexibly. The wing sweeps down (or forward) as in a conventional downstroke, then flips over at the beginning of the backward upstroke to present the same positive angle of attack as on the forward downstroke.

The unique wingbeat of a hovering hummingbird seen from above and from the side. TOP TO BOTTOM The wings sweep forward and then rotate and return in an inverted position. Lift is generated on both forward and backward strokes and drag balances out so that the hummingbird hangs in the air on almost invisible, whirring wings. The fuel for such sustained effort comes from the energy-rich nectar.

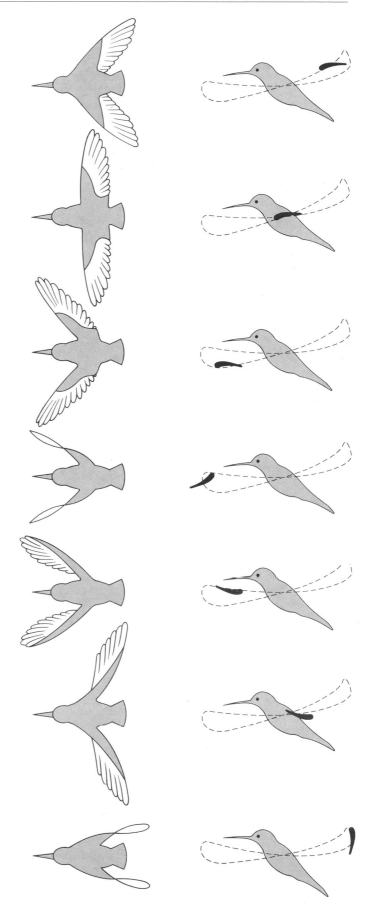

Unusually for a small bird, the wing does not fold on the upstroke. In this way lift is generated continuously and equally on both strokes, with power for the upstroke coming from a relatively large supracoracoideus muscle (over half the size of the pectoralis, compared with only one-fifteenth in a gull). The hummingbird is assisted by the absence of profile and parasite drag at zero airspeed, but on the other hand there is no flow of air over its wings to help generate lift. Induced drag is high, but it is equal on each wingbeat and so balances out.

To manoeuvre, the hummingbird tilts its wings, in the same way as the helicopter pilot alters the angle of the rotors, so that the lifting force is directed forwards or backwards. The result is that a hummingbird can fly forwards or backwards, spin on its axis and even turn over and fly upside down for a moment or two when it is disturbed and needs to escape in a hurry. Some males fly sideways when they display, shuttling to

LEFT A scintillant hummingbird feeding on an epiphytic plant in a cloud forest. In the middle of the backstroke, the wings have turned over and continue to generate lift.

BELOW A stripe-tailed hummingbird steals nectar by piercing the base of a long-necked lily flower.

and fro in front of their mates with their beaks always facing forwards.

To generate enough power for sustained hovering, hummingbirds have large flight muscles which contain dense concentrations of mitochondria (the minute bodies in every tissue cell which are the site of energy release) and are supplied with blood that is rich in oxygen-carrying red cells. The fuel comes from the nectar, essentially an energy-rich sugar solution, although insects and spiders are caught as a supplement. A hummingbird needs about half its body weight of food per day which it can only get by visiting hundreds or thousands of flowers (see page 128).

Some of the sicklebill hummingbirds do not hover while feeding; they cling to the flowers while sipping nectar. This raises the question of why any hummingbird should hover. They would save so much energy if they landed on flowers, like the Australian honeyeaters which usually (but not always) feed on nectar. A possible answer is that honeyeaters feed at plants which have compact bunches of flowers so they can feed from several sources without moving their feet, while hummingbirds have to move from flower to flower. It would take longer if they had to take off and land at each one and feeding would be significantly slower.

CHAPTER 5

LIVING IN THE AIR

Birds use their special ability to good effect. Although some species rarely fly, others are airborne most of their lives, and this chapter examines the ways that birds use their powers of flight for feeding, migration, breeding and to escape from predators. Economy is the rule, but there are times when strenuous activity pays off.

A flock of cape pigeons has gathered near an Antarctic shore where a swarm of marine life floats at the surface.

TIME IN THE AIR

We take it for granted that birds fly, but how do they use this talent that has necessitated such extreme changes in anatomy and physiology? Flying is a very strenuous activity and many birds save energy by their styles of flight. As the table below shows, there is a correlation between the amount of time that a bird spends in the air and its flight costs:

Time spent flying per day compared with the cost of flight, expressed as multiples of the energy used in perching quietly.		
	Hours in air per day	*Flight cost*
Robin	0.5	22.6
Kestrel	1.7	16.2
Starling	2.5	8.1
Swallow	9.1	4.8
Swift	16.6	3.4

Many gamebirds such as pheasants, partridges and grouse, and rails such as coots, moorhens and crakes, rarely fly, preferring to walk most of the time. Some birds are so reluctant to fly that, like the wren-thrush of Central America, they have to be described as 'flightless or nearly so' because, although no one has seen them flying, there is no anatomical reason why they should not. These birds have little use for flight because they feed by looking for tiny morsels on the ground and must move slowly to search thoroughly at every step. The chicks leave the nest and accompany their parents shortly after hatching, so flight is not needed to carry food back to the nest.

The remaining reason for flying is to escape danger, yet some ground-dwelling birds still prefer to run unless very hard-pressed. This does not mean that these birds are incompetent fliers. Consider the corncrake, which is hard to spot because of its skulking habits and usually flies only with great reluctance. There is nothing in its appearance to suggest that it is capable of sustained flight, yet it is a long-distance migrant that travels annually from Europe to tropical Africa. Part of the explanation is that the heavy-bodied corncrake and other rails have wings with high aspect ratios but no wingtip slots. Wings of this sort are suitable for sustained flight but do not allow the bird to take off easily.

On the other hand, vultures, albatrosses, swifts and swallows spend most of their time in the air. Their flight style permits them to travel easily over considerable distances, so they can scour vast areas of sea and land in search of food for their nestlings. Flight is economical for these birds because they glide as much as possible and use soaring techniques to save further energy.

Birds that rarely fly and those that spend much of their time in the air represent both ends of the spectrum. In between there are thousands of species that fly mainly for the three reasons already discussed: to find food, to find a mate and rear their young, and to avoid danger. We can see birds flying all about us and, if we could interrogate them, we would find that nearly every time they were on one of these three important missions.

Very occasionally one sees a bird flying for no apparent reason and the only possible explanation is that the bird is flying for fun. It is quite obvious to anyone who has watched wild birds or kept pets that they sometimes indulge in play, but zoologists have not found it easy to explain, or even define, play. A common sight near a rookery, especially in autumn, is the 'shooting' of the rooks. A flock, maybe a hundred strong, spirals up on the wind with a babble of cawing, and then cascades down, twisting and turning until they reach the treetops. Sir Julian Huxley, one of the pioneers of the study of bird behaviour, described this as a 'sport' and concluded that the rooks were enjoying themselves. It is easy to call an activity play if it looks fun to human eyes and if it serves no obvious purpose, but 'shooting' could be connected in some way with the social life of the rookery.

A bird can only play if it has spare time and energy. Young birds are the most playful, therefore, because their parents have taken on the onerous task of providing food. They also benefit from practising the skills needed for later life. Families of birds of prey make dummy attacks on each other and inanimate objects. Young frigatebirds swoop and seize leaves and other objects floating on the sea, and young kestrels spend hours hovering and pouncing on sticks and pine cones.

It is less obvious why adults play. Why, for instance, do ravens fly up with twigs, drop them and then catch them again in the air? Aerial play often takes place in strong winds or in the currents and updraughts around cliff faces, so the players do not waste too much energy. The strangest 'game' I have heard of involved a pair of white-fronted geese that soared to and fro along a ridge in gale-force winds, occasionally landing briefly or going through the motions of landing. Soaring is not a normal part of goose behaviour, so there seems to be no other explanation than that they enjoyed the unusual opportunity for some effortless flight.

FLIGHT FOR FEEDING

Zoologists studying feeding habits have learned that animals aim to get the best return for their efforts. For birds, economy of flight helps to achieve results, as is admirably demonstrated by the northwestern crow, a species that lives on the Pacific coast of North America, and the European carrion crow. They hunt whelks at low tide and smash their shells by carrying them aloft and dropping them on rocks. Only the largest whelks are selected because they smash easily and yield a worthwhile meal, but the crow takes care not to waste energy by flying too high before dropping a shell. On the other hand, if the crow does not go high enough the shell will not break. Observations show that the crows drop their whelks from about 5 metres (16 feet) which experiments confirm to be the most economical height for breaking shells: any lower and the shells do not break and any higher is a waste of effort.

Economy is a virtue, so if an animal has difficulty finding a meal it begins to search for food somewhere else or changes its

A spotted flycatcher returns to its nest with a butterfly. Large insects can be difficult to chase and catch.

diet. With flight requiring a heavy investment of energy, it is especially necessary to get good results when foraging in the air. A spotted flycatcher, for instance, feeds by flying out from a perch and snapping up passing insects. It prefers medium-sized bluebottles and other flies because each one is a good mouthful. Mosquitoes are so small that they are hardly worth the effort of the chase, unless there is a dense swarm and several can be picked up in a single sally. Bees and butterflies are usually disdained because they are powerful insects and are an effort to chase, kill and dismember.

Birds that feed on the ground have an advantage over aerial species in that they can eat heartily without worrying about the weight of the meal. If a bird has to carry its food in flight, it is hampered by an increased wing-loading. Small birds have enough surplus power to carry a 'payload' equivalent to their own weight. This is an important capability for migrants, but for everyday life it is better to keep the weight down and save on the cost of flight. One explanation for the dawn chorus is that the birds sing at the break of day because, if they fed first, they would be committed to carrying extra weight throughout their waking hours. Similarly, when food is plentiful, kestrels

Above Herring gulls overtake a fulmar encumbered with a heavy fish. They will force it to drop its load.

Left The economy of the common bee-eater is based on flying to catch aerial, and often swift, agile prey.

Below The red-backed shrike hunts like a bird of prey, launching itself from a perch to power-dive on small animals.

store the first catch of the day, so that they do not add to the expense of their hovering flight. In the evening they return to eat the cache and retire to roost with a full crop.

The problem of payload becomes acute with scavengers such as the vultures and the albatross-sized giant petrel of the southern ocean. These birds gather to feed on dead animals and the aim of each bird is to gorge itself by stuffing its crop with as much meat as possible before the food runs out. It then retires from the fray and settles down to digest in peace. When chased, the scavenger runs away and tries to take off but often fails. *In extremis,* it disgorges its hard-won meal to lighten its load before making another attempt at flight.

The need to 'lighten ship' and escape harassment is exploited by birds that seek an easy meal by piracy. The specialists in this field are the frigatebirds and skuas which chase seabirds returning with food for their nestlings. It seems that they can tell when a bird is worth chasing because its flight is laboured and they use their superior flying speed and manoeuvrability to harry it in a mercurial dog-fight, if it is an aerobatic tern, kittiwake or tropicbird, or in a straightforward pursuit and snatch if they have picked on a plodding auk. The only way the victims can escape the nuisance of these aerial buccaneers is to provide them with an easy meal by dropping or disgorging the food that they have laboured to collect, and start foraging afresh.

BIRDS OF PREY

There are two common images of a hunting bird. There is the peregrine, eagle or sparrowhawk hurtling after its victim, and the short-eared owl or kite slowly quartering the ground. Most birds of prey employ one of these methods of hunting. The former are 'attackers' that typically watch from a perch and launch high-speed, agile sorties after passing prey, while the latter are 'searchers' who spend long periods patrolling their hunting grounds in active search of victims.

Attackers have high wing-loadings and this gives them the speed to outpace their prey. Few sights in nature inspire so much emotion as the sight of a falcon stooping on its quarry: the hunter hurtling headlong and the victim twisting and turning to shake off pursuit. Who do you support – the falcon that must kill for a meal or the quarry that must escape? It is a question of survival for both.

In level flight a peregrine may be no faster than a duck or a pigeon, so it attacks by diving, climbing first if necessary. The prey's advantage in agility is offset by the peregrine's weight giving it a near free-fall speed as it swoops down. The peregrine's usual technique is to 'still hunt'. It watches from a perch or circles on the wind; when a quarry appears it gives chase, first spiralling upwards to gain height until it can launch an attack from above and accelerate to overtake the quarry. A mid-air collision at such speed is as potentially damaging to the falcon as to its victim, but a peregrine can dispatch victims larger than itself by dealing them a glancing blow with its feet or slashing them with its hind claws. It may not kill outright; several stoops and blows may be needed before the victim crashes to the ground, leaving the trail of feathers that is the hallmark of a peregrine kill.

Analyses of peregrine kills show that the diet is mainly small to medium-sized birds, from songbirds to pigeons, puffins, partridges and grouse. They can, however, attack and kill herons, barnacle geese and cock capercaillie which are much larger than peregrines. It might be expected that a predator would pursue prey considerably smaller than itself so that it could readily overpower it, but manoeuvrability decreases with increased size, so a predator larger than its prey will have difficulty closing in on a small but more agile victim. Many of the peregrines' victims are fast but not very agile, so they have difficulty evading an attack.

Sparrowhawks specialize in catching small birds so they have to use stealth and surprise. They are best known for their

Below A scops owl about to strike a bush cricket. As it brakes it thrusts its feet forward to increase their momentum and make an effective trap for its prey.

Above After a long search, a hungry buzzard loses height to inspect a possible meal of carrion. It is flying very slowly, with the inner wings stalling.

Right The attacking kestrel has had to strike fast to seize a vole before it can reach the safety of its burrow.

headlong dashes along woodland edges and hedgerows, flying low and slipping through a gap, around a corner or over the top to catch small birds unawares, but this energetic stratagem is not the usual hunting technique. More often, the sparrowhawk flies from perch to perch, pausing at each one for several minutes to scan the surroundings. On spotting prey, it launches an attack, either flying flat out without concealment or making use of cover, depending on how much ground it has to cross before it can strike.

In contrast to the attacking peregrine and sparrowhawk that alternate immobile scanning of the airspace with short, sudden bursts of activity, the searchers spend hours in the air looking for prey. They have low wing-loadings for slow, buoyant, economical flight and they hunt small but abundant prey. Kites and harriers are typical searchers. Long-winged and long-tailed for manoeuvrability at low speed, they range over open ground, gliding easily on outstretched wings between short bursts of flapping. Their prey is caught by surprise rather than pursuit; the searcher drops suddenly out of the air, pouncing before the victim can react.

The impression is often given that a falcon is an infallible 'hunting machine' and that its quarry is doomed from the start. W.H. Hudson considered that the peregrine had 'an

infallible judgement. However swift of wing its quarry might be, it was almost invariably overtaken and struck to the earth; and the bird thus vanquished was in many cases the equal, and sometimes even the superior, in weight to the falcon.'

Some observations suggest that peregrines and merlins miss their target eight or nine times out of ten and the Swedish ornithologist G. Rudebeck recorded only 19 successes out of 252 attacks by peregrines. Ospreys, on the other hand, are reported to be successful in 80 to 90 per cent of their dives after fish, and kestrels have a 40 per cent success rate. There is ample evidence, however, that falcons sometimes play with their quarry, so these figures give a false impression of the efficiency of a falcon that is hunting seriously. The peregrine specialist Derek Ratcliffe has described watching a peregrine sallying from a crag to attack racing pigeons. It singled out a pigeon from each flock, chased it around the hillsides and then lost interest. Not one strike was attempted. I have seen something similar in Greenland where flocks of barnacle geese were migrating down a valley and running the gauntlet of a gyrfalcon. On sighting the falcon, the geese bunched together and flew in circles but often one got left behind, perhaps because its flight feathers had not fully grown after the moult. These geese were attacked repeatedly and forced to the ground, whereupon the gyrfalcon lost interest.

Above The Australian spotted harrier hunts by slowly quartering the ground in search of small animals. The splayed primaries and fanned tail assist slow flight.

Below Birds of prey need split-second timing to catch live animals. This African fish eagle has to adjust its strike on a fish swimming under the surface by making allowance for the refraction of light altering its apparent position.

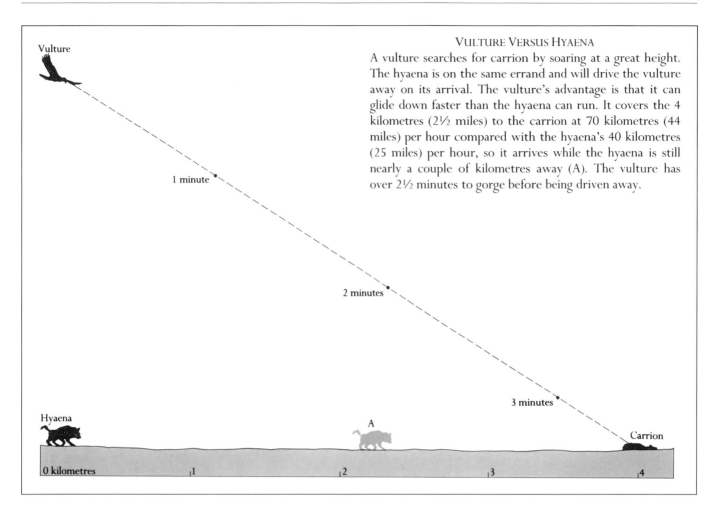

VULTURE VERSUS HYAENA
A vulture searches for carrion by soaring at a great height. The hyaena is on the same errand and will drive the vulture away on its arrival. The vulture's advantage is that it can glide down faster than the hyaena can run. It covers the 4 kilometres (2½ miles) to the carrion at 70 kilometres (44 miles) per hour compared with the hyaena's 40 kilometres (25 miles) per hour, so it arrives while the hyaena is still nearly a couple of kilometres away (A). The vulture has over 2½ minutes to gorge before being driven away.

VULTURES

In *Pirates and Predators*, a classic book of natural history observations, British traveller and naturalist Colonel Meinertzhagen described how he lay in wait after shooting an antelope. After 17 minutes 'a single vulture came swishing down from the heavens.... On looking up I could see birds planing down from all directions, some mere specks in the sky, others half-way down; after another ten minutes there were seventeen vultures on the ground, drawn by seeing others descending.'

Vultures make a powerful image of life and death on the African savannahs, either as flocks soaring overhead on broad, outstretched wings surveying the country below for dead and dying animals, or gathering in ungainly, squabbling mobs to devour a carcass. They also compel interest for the way that they use the swirling air currents over the hills and plains to get the range necessary for exploiting a scattered and unpredictable source of food.

The vultures' technique of staying aloft in the air for hours on end with the minimum expenditure of energy has been studied by Colin Pennycuik. He followed them in a glider and found that there are similarities in the techniques used by glider pilots and vultures because pilots have to keep their machines aloft without any power, while vultures try to use as

little power as possible. Crucial to both is the glide angle, the angle of descent below the horizontal. A small glide angle means that the vulture travels a long way for little loss of height, but the secret of its success at long distance, economical flight lies in its ability to exploit its environment and soar, using rising currents of air to provide lift and save the effort of flapping flight.

Vultures use slope-soaring where the wind is deflected up the face of a hill, and wave-soaring where the wind is thrown into standing waves beyond the hill (see page 100). However, their main method of extracting energy from the atmosphere is thermal soaring.

To travel across country after the migrating herds of game animals and bring a 1 to 2 kilogram (2 to 4½ pound) load of meat back to the nest, the vulture climbs in a thermal until the upward flow of air weakens, then sets off to the next in a series of spiral climbs alternating with long, flat glides. Its progress is helped by thermal 'streets' where a line of thermals is drifting downwind. In such circumstances the vulture may not climb in the thermals but pass straight through them. As it crosses the zone of downward-moving air on the outside of the vortex, it speeds up to minimize the loss of height, then, as it goes through the rising air in the centre of the thermal, it slows down to make the most of the lift.

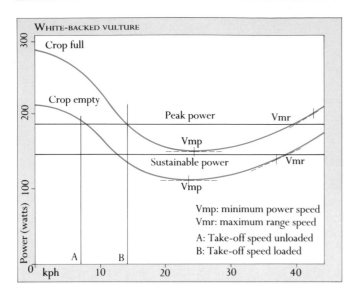

Above: Laden vultures can take off only if they run or use strong thermals, and they barely stay airborne by flapping.

Below: A king vulture finds it easier to take off by leaping from a perch.

Of the vultures living on the African savannahs, the two largest, the white-backed and Rüppell's griffon vultures, are the most adept at cross-country travel. This is because their relatively high wing-loadings give them a fast gliding speed, compared with smaller species such as lappet-faced and white-headed vultures. As we have seen, the disadvantage of high wing-loadings is that take-off in the morning is delayed until the ground has warmed sufficiently to produce thermals strong enough to lift the birds off the ground. However, Rüppell's vulture can get airborne at first light because it nests on cliffs where it can use slope-soaring. White-backed vultures nest in trees and can take off by leaping, but they prefer thermals to develop before setting off across country. Nevertheless, they sometimes set off early, when the sounds from a party of hyaenas on the hunt alerts them to a certain meal that will repay expensive flapping flight.

Cross-country ability is reflected in the hunting habits of the vultures. The smaller species range less widely and fly low over individual hunting territories. In contrast, the two griffon vultures forage over hundreds of square kilometres. They circle at heights of 2,000 metres (6,560 feet) or more, with their heads held low for a clear view of the movements of predators and prey on the ground below, and of other vultures around them. They may spend eight hours on a foraging trip, until bad weather or the setting sun wipes out the thermals.

When a carcass is spotted, the vulture must not waste a moment getting to it. Smaller species, in particular, must snatch a meal before the arrival of larger birds, which have probably been watching overhead. All vultures also have to beat hyaenas to the spot, as they too have been watching the flight paths of descending vultures and will drive off any competition. So, when a meal is sighted, vultures reverse their efforts to maintain height. They fold their wings to reduce the wing area by one third and diminish lift, and lower their legs to treble drag, thus reducing the glide ratio so that gliding angle and speed are increased. The result is the swishing dive witnessed by Colonel Meinertzhagen that gives vultures their competitive edge over hyaenas. The vultures outpace the hyaenas to get a few minutes' advantage, and that is time enough to gorge to repletion.

Provided they are undisturbed, vultures can rest and digest their meals, but breeding birds have to get back to the nest with food for their chicks. Colin Pennycuik caught a white-backed vulture that was unable to take off with 1,140 grams (2½ pounds) of zebra flesh in its crop, so parent vultures are faced with the problem of take-off. They may have to digest some of the food to reduce the load and make use of the strong afternoon thermals to get airborne, but once aloft they make good time back to the nest provided there are strong thermals to carry them up. The extra weight in their crops raises their wing-loading so they glide faster between thermals.

The diagram on this page shows how an unladen white-backed vulture has to run, or leap from a tree, to get up speed for take-off. It can just manage to fly in search of food at its maximum range speed, although it will prefer to fly more slowly and soar. When it has fed, its power-speed curve is shifted to show that take-off is impossible without assistance and it no longer has the power for continuous flight.

HUMMINGBIRDS

The hummingbird's unique hovering flight (see page 115) gives it access to a rich source of energy, but it must ensure that it has enough flowers to satisfy its needs. For the rufous hummingbird that breeds from Oregon to Alaska, this means migrating to southern Mexico for the winter, while tropical species rely on the year-round succession of flowering to provide a continuous supply of blooms in one place.

Although nectar is a rich food, some hummingbirds have to visit as many as 2,700 flowers per day to satisfy their energy requirements. To do this efficiently, hummingbirds spend most of the day perching quietly and only a couple of hours or so feeding. As a rule, large and small species employ different strategies for obtaining their daily nectar. The smaller species usually defend patches of flowers against other hummingbirds. They choose a patch that has just enough flowers because it would be a waste of energy to patrol and chase intruders away from an unnecessarily large area. This was demonstrated by a neat experiment in which some of the flowers in a patch belonging to one hummingbird were enclosed in bags. The bird had to enlarge its patch to take in more flowers, but it reverted to the original size when the bags were removed.

To be effective 'fighters', these territory-holding hummingbirds have small wings and fast wingbeats which give them speed and manoeuvrability. Their hovering flight is consequently expensive but they defend plants with packed masses of flowers, so they do not spend much energy flying from bloom to bloom.

Species with larger wings and slower wingbeats use a strategy known as 'traplining', in which the hummingbird progresses around scattered flowers like the human trapper making the rounds of his line of traps. Hovering is less expensive for these hummingbirds but trapliners fly rapidly between their widely spaced flowers. The long-tailed hermit of Central America flies at 35 to 40 kilometres (22 to 25 miles) per hour between flowers (compared with the territorial cinnamon hummingbird's speed of up to only 4 kilometres/ 2½ miles per hour). This is well above the maximum range speed of 27 kilometres (17 miles) per hour and nearly as expensive as hovering, but travelling at speed means more flowers visited in a feeding bout and a higher intake of nectar.

SWIFTS, SWALLOWS AND MARTINS

The expensive hovering flight of hummingbirds is quite the opposite of the gliding flight of the swifts, swallows and martins, whose long, narrow wings are designed to reduce drag. In general their flight is economical and these birds can spend many hours in the air each day searching for insects. Economical flight is needed for two reasons: they may have to spend some time searching for suitable concentrations of insects and, having found them, flying slowly makes it easier for the birds to pick insects out of the air.

Within the category of species breeding in the British Isles (swallows, house and sand martins, and swifts) flight style is related to diet and breeding. The size of insect each species

ABOVE Broad-tailed hummingbirds feed together. Each hummingbird must have access to enough flowers for its daily energy requirements.

RIGHT The long-tailed hermit flies extra fast between flowers, so that it can visit as many as possible in a feeding bout.

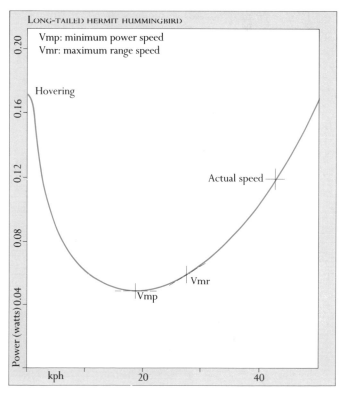

LONG-TAILED HERMIT HUMMINGBIRD

Vmp: minimum power speed
Vmr: maximum range speed

Hovering

Actual speed

Vmr

Vmp

Power (watts)

0.20 0.16 0.12 0.08 0.04

kph   20   40

	Swift	House martin	Sand martin	Swallow
Wing-loading	3.86	4.74	5.03	6.11
Manoeuvrability index	1.8	3.1	3.9	5.2
Arrival time in Britain	Late May	Early May	Early April	Mid-April
Lays eggs	Late May	Mid-May	Late April	Late April
First clutch size	2–3	3–5	4–6	4–6
Percentage of birds laying second clutch	0	87	60–90	67–92

prefers to eat increases in the order of swift, house and sand martins and swallow. This is related not only to the size of the bird's beak but also to its manoeuvrability. In a study of the feeding habits of these species, D.R. Waugh divided the length of the outer part of the tail of each species by its body weight to calculate an 'index of manoeuvrability'. The swallow has an index of 5.2 but the swift's is only 1.8, the sand and house martins being intermediate at 3.9 and 3.1 respectively. These figures accord with the birds' feeding habits.

At the extremes of manoeuvrability there is a clear difference between the swift gliding very economically on a fairly direct course to pick up small, scattered insects, such as aphids which have been swept up in rising air currents, and the swallow flying at a lower level, darting hither and thither more energetically, to seize fast-moving flies. When swallows do feed on aphids, they fly higher than usual, joining the swifts and imitating their habits by gliding more frequently.

The flight patterns of each species also explain the differences in breeding habits, as shown by the table on this page.

Large insects, of the housefly and hoverfly type, appear early in the year, well before warming air lifts aphids off the vegetation. So swallows and sand martins appear in their summer home first, feeding on scarce but large flies, and they breed more prolifically than the late-arriving house martins and swifts.

LEFT Cliff swallows gather over a lake where insects are swarming. Finding a concentration of flying insects makes the search for food much simpler and more rewarding. A prolonged spell of bad weather can bring disaster as the swallows waste energy searching the airspace for a much needed meal.

ABOVE A swallow returns to its nest with its throat bulging with insects for the nestlings. The hunting technique of the swallow enables it to feed on scarce insects, prolonging its stay on the summer breeding ground so that it can raise two or three broods of young before heading south again.

FLYING TO AVOID PREDATORS

One of the first advantages birds got from their power of flight was a speedy escape from enemies and the chance to remove themselves and their families permanently from many dangers by roosting and nesting in trees and on cliff ledges. A tree roost provides security as well as protection from extremes of weather, and ground-dwelling peafowl, pheasants, partridges and turkeys still fly up at dusk to spend the night safe from foxes and other nocturnal ground-predators.

As we have seen, the relationship between predator and prey is not totally one-sided in favour of the predator, since the tactics of birds of prey are not infallible and they sometimes fail through the evasive action of their targets. Once a bird has spotted its assailant the encounter becomes a contest of alertness, speed, agility and stamina.

The hobby is a fast, agile falcon that preys on birds and large insects. Swifts, hirundines and larks are frequent victims.

When Niko Tinbergen studied hobbies on the sandy heaths of central Holland he watched the effect they had on swallows. He noted that some swallows flying above the trees paid no attention to a family of hobbies circling about 300 metres (1,000 feet) higher up, until the latter started playing swooping games. Then the swallows made a mad dash for cover among the trees where the hobbies, if they had attempted a pursuit, would have been at a disadvantage. Similarly, the round-winged, long-tailed sparrowhawk hunts in woods and, although faster than the small songbirds it pursues, can often be outmanoeuvred by them because they are smaller and can dive for cover.

Birds living in flocks in open spaces use the different tactic

BELOW A black-winged stilt chases a much larger herring gull that has come too close to its nest. Speed and agility give the stilt an advantage over the gull in the air.

ABOVE Like other skuas, the long-tailed skua of the Arctic tundra is vigorous in the defence of its nest and young.

of bunching together and circling rather than fleeing. They are like the pioneers of the Wild West who corralled their waggons when they could not outrun the Indians. This is the reaction of starlings, waders, ducks and geese to the appearance of a predatory bird. The flock closes ranks and gives a display of the perfectly co-ordinated evolutions described in Chapter 3. The sight of one of these wheeling flocks is a sure sign that there is a bird of prey in the vicinity.

As far as each individual in the flock is concerned, there is safety in numbers because a predator can catch only a limited number of prey, so the larger the flock the less chance the individual has of being singled out. Another advantage is that the predator may be confused by the wheeling mass of birds. If it attacks, it runs the risk of an accidental collision and injury. The response of a falcon is to make mock attacks until one or more birds, through inferior timing or physical disability, become separated. However, falcons are sometimes bold enough to strike into the mass of birds. Alexander Wetmore, the leading American ornithologist, described seeing a peregrine 'dash through closely massed flocks of flying sandpipers, striking out two or three with as many thrusts of

its claws, allowing each bird to drop and then wheeling swiftly to seize the falling prey before it reached the ground'.

The alternative to the close-formation 'corral' is to upset the predator's attack by behaving erratically. I have flown over coastal saltmarshes and observed waders – such as redshanks, knots and oystercatchers – taking fright and flying away below me in fast zigzags. Any predator trying to intercept them by plotting a closing course is thwarted by its quarry's unpredictable changes in direction. These were the tactics used by ships trying to evade submarines and bombers in the Battle of the Atlantic.

When birds have young to protect they frequently become bold in their defence. Many species make aerial 'dive bombing' attacks on predators, despite the danger to themselves. These include mistle thrushes, swallows, Australian willie wagtails, African drongos which will land on the backs of eagles to peck them, and red-winged blackbirds, blue jays, mockingbirds and thrashers in North America. Swooping attacks are a common habit among the gulls and their relatives the terns and skuas, and many people have been surprised at the vehemence of Arctic terns when they have walked into a colony. They are not large birds but they can draw blood from an exposed scalp. Skuas can be worse because of their large size and I was half-stunned several times while studying the anti-predator

ABOVE A great skua attacking a sheep that has strayed near its nest. This show of aggression is often sufficient to scare the sheep into running away, but skuas sometimes actually land on the sheep to reinforce the message.

RIGHT An avocet chases a marauding black-headed gull. Gulls are inveterate nest-robbers, but they must use speed or stealth to catch the parent birds unawares.

behaviour of the Antarctic brown skua.

The wing or foot of a 1.5-kilogram (3-pound) skua striking across your ear or neck is distinctly unpleasant, but the bird is aiming to miss. It dives with wings held in a high dihedral and its feet lowered to give it control and usually sweeps past before climbing to repeat the near-vertical attack. A direct hit could be dangerous for the skua so its aim, as with other birds that attack intruders near the nest, is simply to distract the predator's attention away from its eggs or chicks.

The ultimate use of flight in protection of the young is to transport them from the scene of danger. A number of species are reported to clutch their young between their legs and fly away with them one at a time. These include the woodcock, spotted sandpiper, willet, redshank, red-tailed hawk and white-browed coucal.

FLIGHT FOR MIGRATION

The migration of birds has been a familiar phenomenon since earliest times. Over two thousand years ago Aristotle wrote that some birds 'migrate, quitting Pontus [on the Black Sea] and the cold countries after the autumnal equinox to avoid the approaching winter, and after the spring equinox migrating from warm lands to cool lands to avoid the coming heat. In some cases they migrate from near at hand; in others they may be said to come from the ends of the world, as in the case of the crane, for these birds migrate from the steppes of Scythia [north of the Black Sea] to the marshlands south of Egypt, where the Nile has its source.'

Despite familiarity with the fact of migration, people were mystified, not only at the birds' mastery of navigation, but at the capacity of such frail creatures to make journeys of thousands of kilometres. But, just as the navigational methods of birds have been elucidated by observation and experiment, so the secrets of their seemingly miraculous powers of endurance have been revealed.

Nearly all birds migrate to some extent but for some the journey is short and they have no special flying ability. Neither is there anything remarkable about the migratory flights of such aerial birds as petrels or swifts. They spend their days in the air, so a long journey is little more than a continuation of their normal activities. There is evidence, however, that oceanic birds still migrate in long hops without feeding. For instance, sooty shearwaters fatten on anchovies off the coast of California in the summer, then return to their nesting places in the southern hemisphere probably without feeding. Nevertheless, the marvel of migration is seen in the many birds that, having spent several sedentary months in one spot, suddenly set off on an intercontinental flight, often non-stop across oceans and deserts.

Some migrants appear to be totally unsuited for long-distance flight. One of these is the European wren, known in North America as the winter wren. It is a woodland bird that lives in the undergrowth and hops down to find insects on the ground. Occasionally it flits across a clearing in a direct flight on rapidly whirring wings like an outsize, brown bumblebee. Many wrens are residents that barely move from season to season, but their continuing presence masks the movements of others. British wrens are joined by continentals who fly across the North Sea, while Swedish wrens travel about 2,500 kilometres (1,560 miles) to southern Spain, and Canadian wrens make a parallel journey to winter by the Gulf of Mexico.

Leaving aside the question of navigation, the two necessities for long-distance flight are efficient design and the capacity to carry enough fuel for the journey. Much of our understanding of these aspects of bird migration has come from the work of hundreds of ornithologists, often amateurs, who catch and ring birds. Ringing, for instance, confirmed that swallows fly from Europe to southern Africa, as shown by this letter from the Bishop of Glasgow and Galloway to the magazine *British Birds* in 1919:

Sir, — I have just returned from South Africa. When in East Griqualand, staying with the Rev. M. Williamson of Ensikeni, Riverside Post, I was shown an aluminium ring that had been taken off a swallow's leg. The bird was picked up about the 21st of February 1919, in Michael Gwinsa's cattle kraal. The ring was marked "Witherby, High Holborn, London", and inside were the letters and figures "J.M.53". The swallow was very thin and exhausted. The natives all thought it boded ill-luck for Michael, and considered that it was a clear case of witchcraft for a bird to appear from nowhere with a ring round its leg and alight in someone's cattle kraal. Archibald Glasgow & Galloway.

Such nuggets of information are now commonplace and an accurate picture can be drawn of the destination and routes of many species. Often the duration of the journey can be discovered as well. At the same time as the bird is ringed, and when it is later recaptured, its body measurements and weight are recorded. One of the measurements is 'wing length' which, in ornithological practice, means the span between the wrist or carpal joint and the tip of the longest primary.

Wing shape, especially of the outer wing, is important for long-distance flight. Small songbirds have rounded wings with a low aspect ratio but a large total area. This gives them a low wing-loading, which allows them to put on weight for migration without the wing-loading becoming impossibly high. The drawback is that the large wing area gives a high profile drag, but small birds overcome this by adopting bounding flight and flying faster than the maximum range speed (see page 97).

Many migrants, however, tend to have more pointed wings which reduce drag. The records of bird-ringers show that the highly migratory sedge and willow warblers, for example, have more pointed wings than non-migratory or sedentary Dartford and Cetti's warblers. Similar comparisons may be made with the round wings of the resident black-headed oriole of India and the long-winged migratory golden oriole, and, in America, the sedentary russet-crowned warbler and the migratory Blackburnian warbler.

There is even variation within single species which have both migratory and sedentary populations. Blackcaps breeding in northern Europe are migrants that winter in Africa and have long pointed wings, while those in southern Europe are resident all year and have shorter, rounded wings. Swedish ornithologists have found that the wingtips of male willow warblers are more pointed than those of the females, which

OVERLEAF *Snow geese, including the dark form known as the blue goose, mass on Lake Manitoba on their migration between Arctic Canada and the southern United States. Geese are noted migrants that fly long distances each year to exploit two different habitats.*

could improve their flight performance. Their explanation is that the males need to migrate faster and arrive at the nesting grounds as early as possible to set up territories. The females' rounded wings are better for flying through dense vegetation in search of nesting material and food for the young.

The most economical way for a bird to migrate is by short hops so that it is not burdened with heavy loads of fat. This strategy is used by birds which stop to feed en route, but a leisurely passage is not possible when there are seas and deserts to cross, particularly in spring when the birds are in a hurry to reach their destinations and start breeding.

Although the object of migration is to travel to escape food shortage, birds are not driven to move by hunger. If they were, it would be too late. Those that do not travel in short hops must first lay down the reserves of fat needed for the journey. This is particularly crucial for birds which perform extra long flights. The longest non-stop flight is the lesser golden plover's 4,000-kilometre (2,500-mile) hop on its way from Alaska to Hawaii and flights of up to 1,000 kilometres (620 miles) are not uncommon for several other species. It seems incredible that small birds can fly so far without stopping, but a bird's range depends on the amount of fuel it can carry relative to its body weight, and not on the absolute amount of fuel.

The whistling swan is a small, fast-flying species that migrates between temperate and Arctic regions in North America. Like geese, these swans fly in V-formation at a great height.

Compared with the larger species, small birds can carry a higher proportion of their weight as fuel, so some songbirds double their weight before migration whereas a goose or swan with such an extra burden of fat would not be able to take off. In practice, this relationship is not so simple since small birds spend relatively more energy on their metabolism when they are resting. Also, because they fly more slowly than large birds and flights take longer, more fuel is used on 'life support' during the journey. So, although small birds can carry proportionally more fuel than a larger bird, they use it at a faster rate, thus the records for long-distance flights are held by the waders: medium-sized birds which strike a balance between fuel storage and fuel consumption.

When the ability of small birds to carry heavy loads of fat is realized, long-distance migration ceases to be a supernormal phenomenon and even the 4-gram (0.14-ounce) ruby-throated hummingbird travelling 800 kilometres (500 miles) across the Gulf of Mexico is seen to be within the bounds of possibility. It accumulates up to 1.5 grams (0.05 ounces) of fat before departure but calculations show that at a cruising speed of 26

Barnacle geese flying over Scotland. When they fly back to the Arctic they must carry enough fat to sustain them on the ocean crossing and ensure survival until the thaw uncovers their food supplies.

kilometres (16 miles) per hour it will use less than half a gram for the crossing in ideal conditions.

The prime requirement for a successful migratory flight is, therefore, the chance to feed well before departure. Given the opportunity, weight is amassed in two or three weeks. Sedge warblers gathering in reedbeds in southern England can put on half a gram a day by feeding on the aphids living in the dying reed stems. A few put on as much as 11 to 18 grams (0.4 to 0.6 ounces) and have enough fuel for a flight lasting around 85 hours, which at 40 to 50 kilometres (25 to 30 miles) per hour would take them to Africa in one hop.

In practice, the migrant is at the mercy of the elements; a hummingbird makes no progress against a headwind of 35 kilometres (22 miles) per hour and unlucky birds disappear into the Gulf of Mexico. Despite the correct design of the wings and a full load of fuel, the bird must enforce strict economies and try to ensure that it travels in the most propitious conditions. The early observers of migration noted that the movements of birds were linked with the wind and, in the first century, the Roman naturalist Pliny wrote of quails:

'. . . when the south wind blows, they never fly.' Sensibly, birds prefer to wait until there are favourable winds for their departure. The influence of the wind is a godsend for British birdwatchers who use weather forecasts to predict the arrival of migrants from Scandinavia. High pressure weather systems over northern Europe result in easterly winds which sweep masses of birds, of every description, across the North Sea to Britain. In October 1965 this weather pattern brought millions of birds to the eastern coast. An estimated half a million birds landed on one 40-kilometre (25-mile) stretch of Suffolk and the town of Lowestoft was swamped with birds, two residents having the unusual experience of redstarts landing on their shoulders. Many birds were exhausted and large numbers of bodies were washed ashore, but when the weather changed the survivors moved on.

On the other side of the Atlantic weather patterns strongly influence the autumn flight of small birds, as every night millions of songbirds and waders leave the coast of North America, between Nova Scotia and Virginia, and disappear out to sea. Their destination is South America and their seemingly suicidal choice of route is explained by the type of weather in which they prefer to travel. The birds wait for a cold front to pass over and take off in the northwest wind that follows. The southeasterly course carries them over Bermuda about 18

hours later and towards the Sargasso Sea. Here, they meet the northeast trade winds and are carried to the coast of South America. The flight, which lasts at least 80 hours, carries them over 3,000 kilometres (1,864 miles) of sea, the longest non-stop flight of any passerine birds, but it is less strenuous than the land route down through Central America because the continuous following wind furnishes the birds with considerable savings in energy.

It is common sense for a bird to wait for a tailwind to blow it along and assist its progress so that it can save energy by coasting. If a bird has an airspeed of 40 kilometres (25 miles) per hour and it is flying into a headwind of 20 kilometres (12 miles) per hour, its groundspeed will be only 20 kilometres (12 miles) per hour. For every kilometre that it flies through the air, it will cover only half a kilometre over the ground. Conversely, the same wind from behind will sweep it along at 60 kilometres (37 miles) per hour.

In theory, the bird should not continue flying at the same airspeed regardless of wind. The power-speed curve (see below) shows that it should slow down and coast, letting the wind carry it. In all previous discussions of power and speed, it has been assumed that the bird is flying in still air. In the diagram the maximum range speed is indicated by a line that runs from the origin of the graph to meet the curve at a tangent. Adding the tailwind shifts the origin to the left, and moves the tangent. The tailwind has increased the bird's groundspeed but to achieve economy, it has reduced its power output and its airspeed. So the migrant bird *travels* faster although it *flies* more slowly and uses less energy. A headwind has the opposite effect of pushing the tangent to the right, up the steeper part of the curve. This demonstrates that in a headwind a bird must fly harder to cover the same distance.

There is evidence from observations of migrants on radar that birds do take account of the wind and adjust their flying

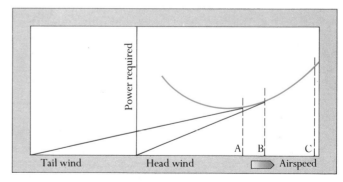

ABOVE A power-speed curve shows that a bird should adjust its speed when flying in a wind. To fly at the maximum range speed (B) it should reduce its airspeed in a tailwind (A) and speed up in a headwind (C)

LEFT It seems amazing that a bird as small as the reed warbler is able to make intercontinental flights. The clue to its astonishing stamina is that it can carry a heavy load of fat in proportion to its body weight.

speeds, as the theoretical power-speed curve indicates, but how they determine windspeed is not known. The theory also indicates that birds should slow down to the minimum power speed when they are lost, to enable them to stay airborne for as long as possible before running out of fuel. Every autumn American birds appear on the coasts of Europe, especially on the western fringes of the British Isles. They have been picked up by westerlies while making the journey to South America and swept across the Atlantic. In these circumstances, it is believed that they abandon the maximum range speed or cruising speed in favour of the minimum power speed and keep flying until they sight land or a passing ship. They can also draw on a reserve source of fuel by metabolizing their flight muscles, breaking down the protein to release energy. Once they have used up their heavy loads of fat, flying requires less power, so they do not need so much muscle.

ECONOMY TRAVEL

There are a number of ways that a migrant can improve its chances of reaching its destination by flying economically. Migrants make use of the flap and glide, bounding and formation flying techniques described in Chapter 4, but for large birds the most important economies are made by soaring, especially in thermals, which leads to further economy because less fuel needs to be carried. Swans and geese are the only large birds that migrate without soaring and they do not have the emarginated wingtips of soaring birds. The only explanations that have been suggested are that it is either more important for them to make the journey quickly, perhaps to reach High Arctic breeding grounds hard on the heels of spring, or that they need to be able to commute daily between their feeding grounds and roosts without wasting precious time.

Soaring reduces the cost of the journey at the expense of speed. Storks migrate through Europe and Asia to Africa in daily hops and have to detour around the Mediterranean, crossing the sea at the narrowest points – the Strait of Gibraltar and the Bosphorus – by spiralling upwards in thermals and gliding across. Cranes following a similar route speed up their migration by flapping across water, and over land when thermals are absent; they also boost their gliding speed between thermals with short bouts of flapping.

There is plenty of evidence from radar studies and observations from aircraft that migrant birds regularly fly at heights of several thousand metres. Radar shows that birds, presumably songbirds, fly over Puerto Rico at 6,800 metres (22,300 feet), an altitude where the air pressure is less than half that at sea level and the temperature is minus 12 degrees centigrade. The benefits of flying in the hostile conditions of high altitude include the avoidance of unpleasant gusting winds, sandstorms and other environmental hazards nearer the ground, and the advantage of stronger and more reliable winds at high level. The whooper swans spotted at 8,200 metres (26,900 feet) by a pilot flying over western Scotland were taking advantage of an airstream blowing them along at

180 kilometres (112 miles) per hour. These altitudes pose no physiological problems because of the birds' unique breathing system, while the freezing air temperatures will, if anything, prevent the birds' bodies overheating from the exertion of the long flight.

Although few experimental studies have been carried out, it is known that wingbeats slow slightly with increasing altitude but they flap more deeply, making flight more strenuous. The aerodynamic advantage is that the air is less dense; this effectively increases the lift/drag ratio, which in turn increases the maximum range speed. In theory, therefore, birds should fly at the greatest height at which they can get enough oxygen to fly at the increased cruising speed. Moreover, windspeed increases with altitude, so a migrant can be swept along with massive energy savings.

A bird with a full load of fat may have difficulty flying and be extra vulnerable to predators, which may explain why so many migrants set off under the cover of darkness. As the flight progresses, fat is burned and the wing-loading drops so flying becomes easier. The bird will be able to reduce its power and speed and climb into thinner air, which enables it to speed up again. This is how a pilot plans a long-distance aircraft flight, ascending as the fuel is burned, and there is evidence that songbirds do in fact gradually ascend during a long flight. There are, however, other constraints. Birds tend to fly lower in headwinds and low cloud and it is possible that they could suffer from icing at high altitudes.

HAWKS ON MIGRATION

The passage of thousands of millions of birds on migration usually goes unnoticed because the birds are flying too high or go past at night. One exception is the movement of birds of prey making use of thermals and other upcurrents. The birds spread out over open country so that they have a better chance of finding a thermal but become concentrated where they can soar along a slope. One of the most spectacular flights can be seen from Hawk Mountain Sanctuary in Pennsylvania, where thousands of broad-winged, red-tailed and sharp-shinned hawks and many other species pass along a line of hills. Concentrations also occur where the birds have to cross water at its narrowest point. They soar to a great height then glide across with as little flapping as possible. In North America the main water crossings take place at Delaware Bay, across Lake Superior, and at San Francisco Bay. In Europe, the migrants can be seen flying over the Strait of Gibraltar, the Bosphorus and the Baltic (from Falsterbo in Sweden across to Denmark).

ARRIVING WITH RESERVES

There is more to migration than simply surviving the journey and reaching the destination alive. In spring, especially, the birds must arrive with sufficient reserves to launch into the breeding season. This is important in the Arctic where the short summer allows no time to be lost. The breeding success of brent geese on Novaya Zemlya was studied by Dutch

The red-tailed hawk is one of the North American birds of prey famous for gathering in impressive numbers where migratory routes funnel the birds along mountain ridges and across water.

ornithologists, even though at the time no Dutchman had been to this Siberian island since Willem Barents discovered it in 1596. They were able to monitor breeding success because the geese fly back to their Dutch wintering home in family parties of adult pair and their brood of goslings. Breeding success is linked, at least in part, to the condition of the females before they set out from the Netherlands in spring. The heavier they are on departure, the more young they bring back in autumn. Weather conditions during the spring migration are also important because tailwinds push the geese north, thereby assisting heavy females. With the wind against her, a female that started out fat could lose so much weight

through the effort of carrying the extra load that she would arrive in poor condition.

It is a sad fact that many birds perish on migration. Observers on ships and shore are familiar with the distressing sight of birds flying ever lower and finally disappearing into the waves. Many more reach the shore but are totally exhausted. Songbirds arriving in good condition with a following wind keep flying inland, but those that have fought a contrary wind to reach land may be so weary that they are almost impossible to flush from the bushes and long grass where they take refuge. Some birds, such as the sharp-shinned hawks that Neal Smith found on an Argentinian shore after their long journey from North America are so weak that they can be picked up by hand. Even if some of these birds survive, it seems unlikely that they will recover condition sufficiently to breed that year.

FLIGHT FOR BREEDING

The imperative to breed and leave descendants is very strong. When biologists talk about an animal being 'successful' they are referring to its achievement at rearing offspring. Raising a family requires a large investment in time and energy and the economy of flight that is practised at other times has to be abandoned. The migration to suitable breeding grounds is, on the broadest scale, part of the investment in breeding, but during nesting there are several activities which require investment in flying time. The male of many species, sometimes joined by the female, performs aerial displays both to a mate and to rivals. Building material has to be carried to the nest site and later food, and sometimes water, has to be carried to the sitting bird and the nestlings.

Aerial displays to establish a territory or court a mate need not involve a great expenditure of energy. Buzzards, for example, circle effortlessly in thermals, but other birds have energetic display flights. There are the vertical leaps of the lesser florican described on page 65 and the long circling flights of the knot (see page 8). The snipe has a 'drumming flight' in which it flies around its territory for up to an hour at a time in a switchback course, alternately climbing and diving. In the dive, the two outer feathers of the snipe's tail swing out and vibrate in the slipstream of the beating wings to make a rapid, hollow thudding sound.

The cost of display is increased when the male bird is burdened with ornamental plumes. These are usually found on birds that display on the ground or on a perch but Jackson's widowbird leaps out of the grass of African savannahs, looking like a flapping bundle of black rags – an effective display that can be seen from a considerable distance. An experiment which involved cutting the tail plumes off one male and gluing them to the tail of another showed that the length of the feathers helps to attract a mate. One must presume that the

BELOW The trailing plumes of the long-tailed widowbird are an advantage in courtship but a disadvantage in flight.

only reason widowbirds have not evolved even longer tails is that they would interfere too much with flight.

In general, long, loose plumes must be a disadvantage, making flight more difficult and exposing the wearer to greater danger from predators. Glover Morill Allen watched a male widowbird fly off with a party of females. The latter were soon out of sight but 'the gorgeous male, with his long plumes undulating behind him like a kite-tail' was left far behind. However, there is one example where long feathers are probably an asset beyond the needs of courtship. As with widowbirds, female swallows prefer males with longer tail streamers. This was again demonstrated by the type of cutting and gluing experiment described above. It showed that males with shorter tails took longer to find a mate because they were frequently rejected. The long-tailed swallows mated earlier, so the females laid earlier which in turn enabled them to lay extra eggs. The end result was that the long-tailed males reared twice as many young as their deprived colleagues. Far from hampering a swallow, a long tail must be an advantage. It does not cause much extra drag in straight flight because the streamers fold together, but when spread it increases

ABOVE The magnificent train of the peacock must make anything but level flight difficult.

BELOW Male great frigatebirds fight for possession of a twig. The winner will carry it to the female for inclusion in the nest.

Above Fulmars do not expend much time or energy on nest-building. The eggs are laid in a shallow depression, sometimes with a few pebbles for a lining.

Left Ospreys build a massive nest of sticks which requires many journeys carrying material. Although old nests are reused, extra material is added each year.

manoeuvrability and so aids the catching of insects.

Little is known about the effect of strenuous displays on the performer. From my experience with the knots it appeared that the males had little time to spend feeding, so it is to be expected that the combination of continuous excercise and lack of food would affect their condition. Nest-building also takes up considerable time and energy and some birds go to extraordinary lengths to fashion a basket for their eggs. Apart from the weight of each load, a large stick or a beakful of grass is an aerodynamic liability and causes immense drag well forward of the centre of gravity.

One of the most difficult loads must be the bundles of wet seaweed that cormorants carry up to their cliff ledges, although Arthur Cleveland Bent, the celebrated American ornithologist, watched double-crested cormorants on St Genevieve Island not only fetching seaweed from just under the colony, but also carrying sticks from the Labrador mainland several kilometres away. The long-tailed tit also

expends a great deal of energy lining its nest with up to 2,500 feathers. It has been calculated that a pair travel up to 1,000 kilometres (620 miles) in the fortnight that it takes to build the nest, but it must be a matter of doubt whether it is worth it for the insulation they provide! The effort is considerable and it is not surprising that the Rev. E.U. Savage's painstaking examination in the 1920s of six long-tailed tit nests showed that the number of feathers in the lining was proportional to the distance of the nest from the poultry-runs which were the source of the feathers.

Compared with nest-building, however, the main strain of breeding would appear to be the transport of food to the nestlings, and in many species including hornbills, birds of prey, tits, finches, gulls and terns, the male has the additional burden of bringing his mate some or all of her food while she is incubating. Many small birds make up to 500 visits to the nest with food each day, and others work almost as hard in their effort to keep their offspring well fed. In these circumstances, the best speed for flying to and from the nest is not the maximum range speed that gives economical flight but some faster speed for delivering enough loads of food to keep the occupants of the nest nourished. Economy is abandoned in favour of breeding success.

Similarly, in winter months kestrels spend more time perch-hunting than wind-hovering, because hovering is a very strenuous form of flight. However, hovering is employed more

frequently in the nesting season because it lets males search open ground that cannot be scanned from a perch and, in summer, supports large numbers of voles. As a result, the kestrels can increase their rate of capturing prey for the incubating females and, later, the nestlings.

It seems reasonable to imagine that a bird putting a lot of effort into carrying food to its young would be likely to suffer, but this is by no means the case. Several small birds are known to lose weight when their eggs hatch and they have to spend a large part of the day gathering food for the nestlings, but the loss may be a strategy for reducing the effort of flight. Starlings fly to and from the nest at 50 kilometres (31 miles) per hour, which is much higher than the maximum range speed of 36 kilometres (22 miles) per hour that they normally use, so a lighter body saves on the cost of flights, especially as the birds are already burdened with a beakful of food.

Two groups of birds that do lose condition during breeding are, paradoxically, those which specialize in economic, soaring flight: vultures and albatrosses. Griffon vultures time their

LEFT The dipper's breeding success depends on rapid delivery of prey from the river to the nest.

RIGHT The male resplendent quetzal's long train is grown for the breeding season but hinders its entry into the nest.

BELOW A fairy tern with fish for its young. Efficient flight is needed by birds that travel long distances to collect food.

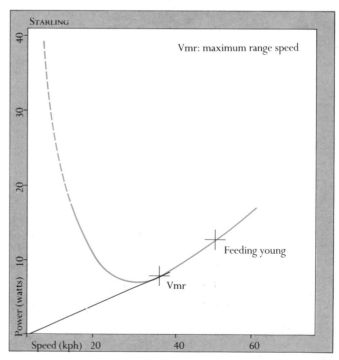

ABOVE When starlings are feeding their nestlings, they fly from feeding ground to nest faster than is economical because they need to deliver the food as quickly as possible.

breeding on the African plains so that there is plenty of food when the young leave the nest, but the movements of the herds of game take them away from the vultures' nesting colonies when the young are in the nest. The adult vultures are forced to travel distances of over 100 kilometres (60 miles) to find enough food for their chicks and they keep going on fat stored before the breeding season.

Albatrosses are also faced with the problem of searching large areas for food and then carrying it back to the nest. The prowess of the ocean-going albatrosses was demonstrated a century ago by a report in *The West Australian* for 21 August 1887 of the finding of a dead albatross, still warm, on a beach near Perth. Around its neck was a tin band punched with the message '*13 naufragés sont refugiés sur les Iles Crozet 4 Août 1887*'. The Crozet Islands were 5,600 kilometres (3,500 miles) away. (Alas, when a rescue party arrived, the shipwrecked sailors had disappeared.) Nowadays the travels of albatrosses are recorded by sophisticated instruments rather than by messages on tin plate. Accurate measurements are made of journey distance and of the time that albatrosses spend bringing food to their nestlings.

Albatrosses nest on oceanic islands where windy days are the norm and cliffs provide strong upcurrents for effortless take-off and landing. They are beautifully adapted for exploiting the almost constant ocean winds but, despite being able to travel with remarkable economy, life does not appear to be very easy for them. The wandering albatross is physiologically capable of breeding at three to four years but it does not attempt to do so for several more summers. Reproduction is then very slow; young are raised only in alternate years and the parents lose weight in the process. They face two problems in gathering food and bringing it back to the nest.

Although the large size – body weight of 8 to 9 kilograms (18 to 20 pounds) and a wingspan of over 3 metres (10 feet) – means that a wandering albatross flies a long distance on a relatively small amount of fuel (85 grams/3 ounces will power it for 1,000 kilometres/620 miles), it has little spare power and may have difficulty in taking off and flying heavily laden with food for its nestling. The unexpectedly small meals the wandering albatross brings for its nestling suggests that this is a limitation. However, it is probably helped by the habit of all albatrosses and their relatives of partly digesting the food that they carry back to the nestlings. The bodies of their prey are partly reduced to oil. The main reason may be to provide the bird with water, but reducing the payload must be advantageous.

Another problem for albatrosses is that of finding food in the huge expanse of the ocean. The grey-headed albatross, smaller but similar to the wandering albatross, spends only a quarter of each foraging trip on the surface of the water, catching mainly squid and fish. Each parent delivers a meal to the nestling on average every two and a half days and the time between each visit is spent travelling up to 500 kilometres (300 miles) from the nest.

FIRST FLIGHT

In the same summer that I watched the song-flights of the knots, I had the opportunity to watch the departure of fledgling snow buntings from their nests. The nests were hidden under piles of rocks and, as the chicks became bolder, they came to the entrance to await the return of their parents with beakfuls of insects. Eventually, the mother (I never saw the male behave in this way) would refuse to push her load into a gaping beak and hopped or flew away, drawing the chick from the safety of the nest and often inducing it to fly after her for up to 200 metres (650 feet) before she relented and landed to let the young bird receive its food.

I was struck by the confidence with which the nestlings launched themselves into the air and flew after their mothers. The power of flight has to be instinctive because it is not possible to learn the complicated movements of flapping flight gradually. It is like being thrown in at the deep end; if you do not know how to swim already, you will drown. The instinctive nature of flight was demonstrated in 1938 by a German scientist, J. Grohmann, who bred two sets of pigeon nestlings: one set was raised in tubes so that they could not

This juvenile wandering albatross is too young to leave its nest, but it exercises its long wings and will soon be flapping hard enough to lift itself off the ground, although this 'practice' is not necessary for the first flight to be successful.

spread their wings, the others lived in nests where they could exercise by flapping. Eventually, both sets were taken out and thrown into the air, and there was no difference in their flight.

It is, nevertheless, easy to believe that birds are taught to fly or at least take time to learn how to do it. Some common species, such as tits and thrushes, seldom fly far for the first few days after leaving the nest. At most they make short journeys from bush to bush in a rather clumsy fashion. It looks as if they are learning how to use their wings, but their problem is that they are still growing their wing and tail feathers so flight is laboured and awkward. The advantage of leaving the nest early is that the brood spreads out for safety's sake, but the young birds will continue to receive food from their parents until feather growth is complete.

ABOVE A juvenile Laysan albatross, still with down on its head, gets airborne on its first flight.

RIGHT A complete change in life for a juvenile little owl. After 30 days confined to the nest, it launches itself into the air and enters a wider world of three dimensions.

The behaviour of house martins has given rise to many stories of birds teaching their young. As the time to leave the nest approaches, the young martins peer out as their parents, often joined by other adults, fly slowly past calling to the youngsters. Eventually one parent, usually the mother, lands beside the nest, then flies off again with the young following close behind. Usually they return to the nest after a short flight, but one sometimes lands on the ground and the adults swoop low over it, apparently trying to persuade it into the air once more.

For some time after they can fly young martins receive food from their parents, supplementing what they can catch until they are fully able to fend for themselves. The swift, whose aerial feeding habits are so similar to the martin's, misses even this assistance in its transition to independence. Once it has left the nest it is on its own. It not only has the ability to fly but is also endowed with all the skills needed to catch its food.

For some birds, premature departure from the nest is the rule. Young guillemots and razorbills brought up on narrow cliff ledges flutter down to the sea before their wing feathers have grown properly. This relieves their parents of bringing food back from far afield, which must be a strain for birds with small wings and laborious flight even when unladen. The difficulty of feeding the nestling is increased by its vulnerabil-

ity to marauding gulls which means that one parent has always to stand guard. (It also risks being knocked off the ledge by the rowdy behaviour of the adults.) Once in the water the young bird accompanies and is fed by its father for several weeks until its wing feathers have grown and it can fly properly.

Young gannets are independent when they leave the nest, but they are so fat that they glide down to the sea and have to fast before they can take off again. Albatrosses, fulmars and their relatives also take off by themselves but they can fly well because they have lost weight and have completed the growth of their flight feathers.

While the basic technique of flapping flight requires no practice, the young bird may take some time to develop its aerobatic skills, including take-off and landing. For this reason birds of prey often stay within their parents' territories for many weeks, perhaps until the next breeding season, and continue to receive food to eke out their own catches.

Falconers know that eyasses (birds taken from the nest) never fly as skilfully as haggards (birds taken as adults from the wild) which have had to learn to kill to survive.

In a study of black kites in the Coto Doñana of southern Spain, juveniles were watched as they practised their flying skills during the two to five weeks' dependence on their parents after they have left the nest. Their first flights are straight flaps from one perch to another. Rather surprisingly they do not glide, which would seem much easier, but the flight feathers are not fully grown until two or three weeks after the kites start flying and wing-loading may be too high to glide satisfactorily. However, once they can glide well, the young birds start to soar. Another surprise finding is that kites do not attempt to hunt during the dependent period and rely totally on their parents. They are then faced very suddenly with the need to find food for themselves.

Niko Tinbergen found a very different situation with young hobbies. Their first flights were clumsy, especially when it came to landing, but they were soon able to take food from their parents in mid-flight, although they were awkward and often had to make several attempts. Within a week they were chasing and catching flying beetles and later their victims included dragonflies. To practise hunting birds, the hobbies made mock attacks on brothers and sisters, swooping down and extending a foot as they passed, to the consternation of nearby swallows (see page 132), until they were ready to take on a real target.

It is interesting that young peregrines, sharp-shinned hawks and goshawks are lighter than adults and so have lower wing-loadings. They also have longer tails. This gives them greater manoeuvrability for catching prey, while also reducing the amount of food needed for survival, but their lack of weight diminishes their striking force. Once they have grown up and become skilled at hunting, their increased weight and speed become an advantage for chasing fast and robust prey.

From the evidence of the pigeons reared in tubes and wild birds growing up in cramped burrows and crevices, it is clear that birds fly by instinct. This is one of the wonders of the natural world; we cannot even walk without hours of practice. Yet it is essential that birds should have evolved this instinctive ability because they do not have the luxury of time to learn to fly; life is difficult enough for any young animal as it launches itself into the world without having to spend time on the basics of locomotion. On the other hand, it is not surprising to find that young birds, especially those that use precision flying to catch their food, need time to bring their flying skills to perfection. We can compare this with the human pilot who must graduate from basic training to take responsibility for an airliner or learn to manoeuvre a fighter plane. As with so many aspects of birds' mastery of the air, however, our understanding of the learning process is far from complete, but we have found out enough to appreciate the advantages that flying brings to birds and to perpetuate mankind's long-standing envy of bird flight.

LIST OF SCIENTIFIC NAMES

A list follows of English and scientific names of birds mentioned in the text. The scientific names are compiled in the order given in *A Checklist of the Birds of the World* by E.S. Gruson (Collins, 1976). Order and families are arranged according to *A Dictionary of Birds* edited by B. Campbell and E. Lack (T. & A.D. Poyser, 1985).

Order Struthioniformes

Struthionidae	ostrich	*Struthio camelus*
Dromaiidae	emu	*Dromaius novaehollandiae*

Order Procellariiformes

Diomedeidae	grey-headed albatross	*Diomedea chrysostoma*
	royal albatross	*D. epomophora*
	wandering albatross	*D. exulans*
	sooty albatross	*D. fusca*
	Laysan albatross	*D. immutabilis*
	black-browed albatross	*D. melanoprhis*
Proellariidae	giant petrel	*Macronectes giganteus*
	cape pigeon	*Daption capense*
	fulmar	*Fulmarus glacialis*
	sooty shearwater	*Puffinus griseus*
Hydrobatidae	Wilson's storm petrel	*Oceanites oceanicus*

Order Pelecaniformes

Phalacrocoracidae	king cormorant	*Phalacrocorax albiventer*
	shag	*P. aristotelis*
	blue-eyed shag	*P. atriceps*
	double-crested cormorant	*P. auritus*
Sulidae	gannet	*Sula bassana*
	cape gannet	*S. capensis*
	blue-footed booby	*S. nebouxii*
Pelecanidae	brown pelican	*Pelecanus occidentalis*
	white pelican	*P. onocrotalus*
Fregatidae	magnificent frigatebird	*Fregata magnificens*
	great frigatebird	*F. minor*

Order Ciconiiformes

Ardeidae	grey heron	*Ardea cinerea*
Ciconiidae	painted stork	*Ibis leucocephalus*
	white stork	*Ciconia ciconia*
	marabou stork	*Leptoptilos dubius*
Threskiornithidae	African spoonbill	*Platalea alba*

Order Anseriformes

Anatidae	whistling swan	*Cygnus columbianus*
	Bewick's swan	*C. bewickii*
	whooper swan	*C. cygnus*
	trumpeter swan	*C. buccinator*
	mute swan	*C. olor*
	white-fronted goose	*Anser albifrons*
	greylag goose	*A. anser*
	snow goose	*A. caerulescens*
	bar-headed goose	*A. indicus*
	brent goose	*Branta bernicula*
	Canada goose	*B. canadensis*
	barnacle goose	*B. leucopsis*
	mallard	*Anas platyrhynchos*
	eider	*Somateria mollissima*
	harlequin duck	*Histrionicus histrionicus*
	bufflehead	*Bucephala albeola*

Order Cathartiformes

Cathartidae	turkey vulture	*Cathartes aura*
	black vulture	*Coragyps atratus*
	king vulture	*Sarcorhamphus papa*
	Californian condor	*Vultur californianus*
	Andean condor	*V. gryphus*

Order Accipitriformes

Accipitridae	swallow-tailed kite	*Elanoides forficatus*
	black-shouldered kite	*Elanus notatus*
	black kite	*Milvus migrans*
	bald eagle	*Haliaeetus leucocephalus*
	African fish eagle	*H. vocifer*
	griffon vulture	*Gyps fulvus*
	Ruppell's griffon vulture	*G. ruppellii*
	lappet-faced vulture	*Aegypius tracheliotus*
	white-backed vulture	*Gyps africanus*
	short-toed eagle	*Circaetus gallicus*
	bateleur eagle	*Terathopius ecaudatus*
	Spotted harrier	*Circus assimilis*
	goshawk	*Accipiter gentilis*
	sparrowhawk	*A. nisus*
	sharp-shinned hawk	*A. striatus*
	buzzard	*Buteo buteo*
	red-tailed hawk	*B. jamaicensis*
	broad-winged hawk	*B. platypterus*
	Swainson's hawk	*B. swainsoni*
Pandionidae	osprey	*Pandion haliaetus*

Order Falconiformes

Falconidae	merlin	*Falco columbarius*
	Eleanora's falcon	*F. eleonorae*
	peregrine	*F. peregrinus*
	bat falcon	*F. rufigularis*
	gyrfalcon	*F. rusticolus*
	hobby	*F. subbuteo*
	kestrel	*F. tinnunculus*

Order Galliformes

Phasianidae	capercaillie	*Tetrao urogallus*
	grouse	*Lagopus lagopus*
	ptarmigan	*L. mutus*
	partridge	*Perdix perdix*
	quail	*Coturnix coturnix*
	chicken	*Gallus gallus*
	pheasant	*Phasianus colchicus*
	peafowl	*Pavo cristatus*
	turkey	*Meleagris gallopavo*
Opisthocomidae	hoatzin	*Opisthocomus hoazin*

Order Gruiformes

Rallidae	corncrake	*Crex crex*
	moorhen	*Gallinula chloropus*
	coot	*Fulica atra*
Gruidae	crane	*Grus grus*
Otididae	great bustard	*Otis tarda*
	kori bustard	*Ardeotis kori*
	lesser florican	*Sypheotides indica*

Order Charadriiformes

Haematopodidae	oystercatcher	*Haematopus ostralegus*
Recurvirostridae	avocet	*Recurvirostra avosetta*
	black-winged stilt	*Himantopus himantopus*
Charadriidae	golden plover	*Pluvialis apricaria*
Scolopacidae	bar-tailed godwit	*Limosa lapponica*
	spotted sandpiper	*Tringa macularia*
	redshank	*T. totanus*
	willet	*Catoptrophorus semipalmatus*
	snipe	*Gallinago gallinago*
	woodcock	*Scolopex rusticola*
	dunlin	*Calidris alpinus*
	knot	*C. canutus*

Stercorariidae	long-tailed skua	*Stercorarius longicaudus*
	great skua	*S. skua*
	brown skua	*S. lonnbergi*
Laridae	herring gull	*Larus argentatus*
	laughing gull	*L. atricilla*
	black-headed gull	*L. ridibundus*
	kittiwake	*L. tridactylus*
Sternidae	South American tern	*Sterna hirundacea*
	fairy tern	*S. nereis*
	Arctic tern	*S. paradisaea*
Rynchopidae	black skimmer	*Rynchops nigra*
Alcidae	great auk	*Alca impennis*
	little auk	*Alle alle*
	razorbill	*Alca torda*
	guillemot	*Uria aalge*
	puffin	*Fratercula arctica*

Order Apodiformes

Apodidae	white-collared swift	*Cypseloides zonaris*
	chimney swift	*Chaetura pelagica*
	European swift	*Apus apus*
Trochilidae	ruby-throated hummingbird	*Archilocus colubris*
	stripe-tailed hummingbird	*Eupherusa eximia*
	long-tailed hermit hummingbird	*Phaethornis superciliosus*
	giant hummingbird	*Patagona gigas*
	amethyst woodstar hummingbird	*Calliphlox amethystina*
	rufous hummingbird	*Selaphorus rufus*
	broad-tailed hummingbird	*S. platycercus*
	scintillant hummingbird	*S. scintilla*
	sicklebill hummingbird	*Eutoxeres aquila*

Order Columbiiformes

Columbidae	snow pigeon	*Columba leuconota*
	domestic pigeon	*C. livia*
	woodpigeon	*C. palumbus*
	collared dove	*Streptopelia decaocto*
Raphidae	dodo	*Raphus cucullatus*

Order Psittaciformes

Psittacidae	cockatiel	*Nymphicus hollandicus*
	crimson rosella	*Platycercus elegans*
	budgerigar	*Melopsittacus undulatus*

Order Cuculiformes

Cuculidae	white-browed coucal	*Centropus superciliosus*

Order Strigiformes

Tytonidae	barn owl	*Tyto alba*
Strigidae	scops owl	*Otus scops*
	snowy owl	*Nyctea scandiaca*
	little owl	*Athene noctua*
	tawny owl	*Strix aluco*

Order Trogoniformes

Trogonidae	resplendent quetzal	*Pharomachrus mocino*

Order Coraciiformes

Alcedinidae	kingfisher	*Alcedo atthis*
	pied kingfisher	*Ceryle rudis*
Meropidae	bee-eater	*Merops apiaster*

Bucerotidae	great pied hornbill	*Buceros bicornis*
	ground hornbill	*Bucorvus leadbeateri*

Order Piciformes

Picidae	green woodpecker	*Picus viridis*

Order Passeriformes

Alaudidae	skylark	*Alauda arvensis*
Hirundinidae	cliff swallow	*Hirundo flavicola*
	swallow	*H. rustica*
	sand martin	*Riparia riparia*
	house martin	*Delichon urbica*
Cinclidae	dipper	*Cinclus cinclus*
Troglodytidae	wren	*Troglodytes troglodytes*
Mimidae	catbird	*Dumetella carolinensis*
	mockingbird	*Mimus polyglottos*
Turdidae	wren thrush	*Zeledonia coronata*
	robin	*Erithacus rubecula*
	redstart	*Phoenicurus phoenicurus*
	blackbird	*Turdus merula*
	mistle thrush	*T. viscivorus*
Sylviidae	Cetti's warbler	*Cettia cetti*
	sedge warbler	*Acrocephalus schoenobaenus*
	reed warbler	*A. scirpaceus*
	blackcap	*Sylvia atricapilla*
	Dartford warbler	*S. undata*
	willow warbler	*Phylloscopus trochilus*
	goldcrest	*Regulus regulus*
Muscicapidae	pied flycatcher	*Fidecula hypoleuca*
	spotted flycatcher	*Muscicapa striata*
	willie wagtail	*Rhipidura leucophrys*
Aegithalidae	long-tailed tit	*Aegithalos caudatus*
Paridae	coal tit	*Parus ater*
	blue tit	*P. caerulescens*
	great tit	*P. major*
Sittidae	nuthatch	*Sitta europaea*
Oriolidae	golden oriole	*Oriolus oriolus*
	black-headed oriole	*O. xanthornus*
Landiidae	red-backed shrike	*Lanius collurio*
Corvidae	blue jay	*Cyanocitta cristata*
	jay	*Garrulus glandarius*
	magpie	*Pica pica*
	chough	*Pyrrhocorax pyrrhocorax*
	carrion crow	*Corvus corone*
	northwestern crow	*C. caurinus*
	raven	*C. corax*
	rook	*C. frugilegus*
	jackdaw	*C. monedula*
Sturnidae	glossy starling	*Aplonis panayensis*
	shining starling	*A. metallicus*
	starling	*Sturnus vulgaris*
Passeridae	house sparrow	*Passer domesticus*
Ploceidae	quelea	*Quelea quelea*
Viduidae	Jackson's widowbird	*Euplectes jacksoni*
	long-tailed widowbird	*E. progne*
Fringillidae	chaffinch	*Fringilla coelebs*
Parulidae	Blackburnian warbler	*Dendroica fusca*
	russet-crowned warbler	*Basileuterus coronatus*
Emberizidae	snow bunting	*Plectrophenax nivalis*
Icteridae	red-winged blackbird	*Agelaius phoenicus*

FURTHER READING

For a non-technical introduction to bird flight and related topics I recommend the following books:

Elkins, Norman. *Weather and Bird Behaviour*, T. & A.D. Poyser, 1983

Gray, James. *Animal Locomotion*, Weidenfeld & Nicolson, 1968

Kerlinger, Paul. *Flight Strategies of Migrating Hawks*, University of Chicago Press, 1989

Mead, Chris. *Bird Migration*, Country Life, 1983

Ratcliffe, Derek. *The Peregrine Falcon*, T. & A.D. Poyser, 1980

Schmidt-Nielsen, Knut. *How Animals Work*, Cambridge University Press, 1972

Village, A. *The Kestrel*, T. & A.D. Poyser, 1990

For the more technically minded the following books probe deeper into the aerodynamics of flight:

Childress, S. *Mechanics of Swimming and Flying*, Cambridge University Press, 1981

Pennycuik, Colin J. *Animal Flight*, Edward Arnold, 1972

Pennycuik, C.J. *Bird Flight Performance*, Oxford University Press, 1989

Ward-Smith, A.J. *Biophysical Aerodynamics and the Natural Environment*, John Wiley & Sons, 1984

The following list of papers published in scientific journals is not comprehensive but gives an introduction to the literature on bird flight:

East, M. 1980. Time budgeting by European robins: inter and intrasexual comparisons during autumn, winter and early spring. *Ornis Scandinavica* **13**:85-93

Ewald, P.W. and Carpenter, F.L. 1978. Territorial response to energy manipulations in the Anna hummingbird. *Oecologia* **31**:277-292

Hails, C.J. 1979. A comparison of flight energetics in hirundines and other birds. *Comparative Biochemistry and Physiology* **63A**:581-585

Hecht, M.K. (Ed.) 1985. The beginning of birds. *Proceedings of the International Archaeopteryx Conference*, Eichstatt, 1984

Hummel, D. 1980. The aerodynamic characteristics of slotted wing-tips in soaring birds. *Proceedings of the 17th International Ornithological Congress*, Verlag der Deutschen Ornithologen-Gesellschaft 391-396

Kodric-Brown, A. and Brown, J.H. 1978. Influence of economics, interspecific competition and sexual dimorphism on territoriality of migrant rufous humming-birds. *Ecology* **59**:285-296

Lissaman, P.B.S. and Shellenberger, C.A. 1970. Formation flight of birds. *Science* **168**:1003-1005

Martin, L.D. 1983. The origin of birds and of avian flight. In *Current Ornithology*, Johnston, R.F. (Ed.) **1**:105-126 Plenum Press

McGowan, C. 1989. Feather structure in flightless birds and its bearing on the question of the origin of feathers. *Journal of Zoology* **218**:537-548

Norberg, U.M. 1981. Flight, morphology and the ecological niche in some birds and bats. *Symposia of the Zoological Society of London* **48**:173-197

Norberg, U.M. 1981. Optimal flight speeds in birds when feeding young. *Journal of Animal Ecology* **50**:473-477

Pedley, T.J. (Ed.) 1977. Scale effects in animal locomotion. Academic Press. (Papers by several authorities.)

Pennycuik, C.J. 1968. Power requirements for horizontal flight in the pigeon. *Journal of Animal Ecology* **49**:527-555

Pennycuik, C.J. 1969. The mechanics of bird migration. *Ibis* **111**:525-556

Pennycuik, C.J. 1972. Soaring behaviour and performances of some East African birds from a motor glide. *Ibis* **114**:178-218

Pennycuik, C.J. 1979. Energy costs of locomotion and the concept of foraging radius. In *Serengeti: dynamics of an ecosystem*, Sinclair, A.R.E. and Norton-Griffiths, M. (Eds.), University of Chicago Press 164-184

Pennycuik, C.J. 1975. Mechanics of flight. In *Avian Biology* **5**:1-73, Farner, D.S. and King, J.R. (Eds.) Academic Press

Pennycuik, C.J. 1982. The flight of petrels and albatrosses. *Philosophical Transactions of the Royal Society of London* **300**:75-106

Pennycuik, C.J. 1987. Flight of seabirds. In *Seabirds: feeding ecology and role in marine ecosystems*, Croxall, J. (Ed.), Cambridge University Press 43-62

Pennycuik, C.J., Alerstam, T, and Larsen, B. 1979. Soaring migration of the common crane observed by radar and from an aircraft. *Ornis Scandinavica* **10**:241-251

Rayner, J.M.V. 1982. Avian flight energetics. *Annual Review of Physiology* **44**:109-119

Simpson, S.F. 1983. The flight mechanism of the pigeon *Columbia livia* during take-off. *Journal of Zoology* **200**:435-443

Torre-Burno, J.R. and LaRochelle, J. 1978. Respiratory exchange and evaporative water loss in the flying budgerigar. *Journal of Experimental Biology* **48**:67-87

Westerterp, K.R. and Bryant, D.M. 1984. Energetics of free existence in swallows and martins during breeding. *Oecologia* **62**:376-381

Westerterp, K.R. and Drent, R.H. 1985. Energetic costs and energy saving mechanisms in parental care. In *Proceedings of the 18th International Ornithological Congress* **1**:392-398, University of Ottawa Press

Withers, P.C. and Timko, P.L. 1977. The significance of ground effect to the aerodynamics of flight and energetics of the black skimmer (*Rhyncops nigra*). *Journal of Experimental Biology* **70**:13-26

Withers, P.C. 1979 Aerodynamics and hydrodynamics of the 'hovering' flight of Wilson's Storm Petrel. *Journal of Experimental Biology* **80**:83-91

Withers, P.C. 1981. Wing tips, slots and aerodynamics. *Ibis* **123**:239-247

Wolf, L.L and Hainsworth, F.R. Time and energy budgets of territorial hummingbirds. *Ecology* **52**:980-988

INDEX

A

aerodynamics 11, *15*, 21, 24, *24*, *25*, *26*, *27*, 32, *35*, 47, 55, 90, 97, 111, 142, 147
aerofoil *19*, 24, *24*, *25*, *26*, 28, *29*, 30, 36, 38, 70, 80
airbrakes *33*, 70, 73, *74*, *124*
airsacs 50, 54, *54*, 55
albatross 9, 11, 28, *33*, 58, 69, 75-76, 90, 98, 101, 105, *106*, 107, 109, 114, 120, 149, 150, 152
   black-browed 69, *82*, *105*, *106*, *109*
   grey-headed 150
   Laysan *152*
   royal 9
   sooty *33*
   wandering 48, *60*, 92, *106*, 150, *151*
alula 28, *29*, *32*, 33, *35*, 36, 69, 70, 74
angle of attack *24*, *25*, 26, *26*, 32, *32*, 33, 36, *36*, 37, 70, 74, *74*, 75, 80, 82, 113, 115
Archaeopteryx lithographica 15-18, *15*, *17*, 50, 79, 83
Argentavis magnificens 50
aspect ratio 26, *27*, 36, 37, 69, 90, 92, 95, *95*, 97, 103, 120, 135
auk 37, 59, 100, 109, 123
   great 42
   little 59
avocet *134*

B

banking 80, *80*, 84
barbet 52
barbs *30*, 31, *31*, 57, 58
bee-eater 82, 90, 96, *123*
Bernoulli effect 24, *25*
blackbird 113
   red-winged 86, *87*, 133
blackcap 135
blood circulation 54, *54*, 55
booby, blue-footed 76
brain 15, 16, 79, 83, 84
breastbone 15, 16, *17*, 18, *51*, 52, *52*
breathing 50, 54, *54*, 142
breeding 142-143, 144-154
budgerigar 50, 80, *86*
bufflehead *73*
bunting, snow 150-151
bustard 52, 65
   great 65
   kori 42, *42*, 65
   lesser florican 65-66, 144
buzzard 44, *53*, 78, 105, 113, *125*, 144

C

capercaillie 124
carpal joint 28, *28*, 30, *30*, 40, *40*, 62, 115, 135
carpometacarpus 28, 30
cassowary 52
catbird 43
centre of gravity 28, 50, 52, *80*, 147
centre of pressure 80, *80*
chaffinch 54
chicken 53
chord 26, *27*
chough 98
clap-and-fling 62, 65
cockatiel *70*
Compsognathus 15, 16
condor 50, 65, 103
   Andean 65
   Californian 42, 47, *47*, 102
coot 16, 75, 120
cormorant 100, 109, 147
   double-crested 147

   king *43*
corncrake 120
coucal, white-browed 134
courtship 82, 144, *144*, 145
crake 120
crane 36, 48, 59, 90, 96, 111, 142
crimson rosella 57
crow
   carrion 24, 38, 48, 55, 96, 121
   northwestern 121

D

diet 50, 52, 53
dihedral *78*, 79, 134
dipper *92*, *149*
display 144, 145, 147
diver 37, 59, 69, 75, 109
dodo 42
dove, collared 36
drag 24, *24*, 25-26, *26*, 28, 31, 32, 38, 43, *43*, 44, *44*, 62, 65, 70, 73, 82, 90, 105, 111, 128, 129, 145, 147
   induced 25, 36, 90, 97, 109, 117
   parasite 26, 117
   profile 26, 50, 97, 117, 135
drongo 133
duck 36, 43, 59, 90, 124, 133
   eider 48
   harlequin 65
dunlin 59
dynamic soaring 105-107

E

eagle 58, 96, 124, 133
   bateleur 57, 82
   African fish *126*
   bald *32*, 36, *83*
   short-toed 114
egg-laying 50, 59, 130
emu 52
energy 38, 42, 43, 47, 49, 52, 53, 58, 66, 90, 96, 97, *97*, 102, 109, *115*, 117, 120, 121, 129, 138, 141, 142, 144, 147
   budget 42
eyesight 83, 84, 86

F

falcon 36, 58, 90, 92, 124, 125, 126, 132, 133
bat 74
   Eleanora's 100-101
feather 14, 15, 16, 30-31, 55
   contour *29*
   covert 70
   evolution of 16, 30
   flight 15, *15*, 28, *28*, *29*, 30, *30*, *31*, *41*, 57, 59, 65, 152, 154
   growth of 151, 152, 154
   ornamental 144-145
   primary 16, 19, 28, *29*, 30, 33, 36, *36*, 38, 40, *41*, 50, 58, 59, *59*, 62, *62*, 66, 70, *74*, *126*, 135
   secondary 16, 28, *29*, 30, 33, 40, 82
   structure *30*
   vaned 31, 36
feeding 83, *149*, 120, 121-130, 138, 139, 142, 147, 150, 152
finch 97, 147
flight
   adaptation to 50, *51*, 120
   bounding 97, *97*, 135, 142
   control 77-87
   evolution of 14-19, 50, 83

   first 150-154
   flapping 16, 21, 28, 38, 40, 42, 43, 47, *47*, 48, 50, 57, 90, 96, 96, 97, *97*, 98, 102, 103, 109, 115, 142, 151, 152
   formation 86-87, *87*, *110*, 111, *111*, *138*, 142
   gliding 8, 14, *15*, 16, 21, 28, 32-33, *33*, 36, *36*, 74, *78*, 80, *80*, *82*, 90, 96, *96*, 97, *97*, 98, 100, 102, *108*, 109, *111*, *114*, 120, 128, 142, 154
   hovering 43, 47, *47*, 49, 69, 80, 100, 113-117, *113*, *114*, *115*, 123, 129, 147, 149
   muscles 16, *17*, 18, *45*, 48, 50, *51*, 52, *52*, *53*, 117
   soaring 8, 47, 48, 50, 57, 80, 98-107, *100*, *101*, *103*, *104*, *106*, 107, 120, 142, 149, 154
   time spent in 120
   undulating 96, *96*, 97
flightless birds *15*, 42, 52, 120
flock 86-87, 111, 120, 126, 133
flycatcher 11, 92
   pied 92
   spotted 121, *121*
frigatebird 27, 37, 50, 69, 79, 82, 90, 92, 103, 120, 123
   great *145*
   magnificent *78*
fulmar *27*, *35*, 48, 75, 80, 96, 100, *101*, *123*, 147, 152
furcula 15, 16, *17*, 52, 55

G

gamebird 42, 53, 66, 120
gannet *28*, 76, 84, *84*, 86, *98*, 107, 152
   cape *21*
glide angle 32, 33, 73, 127, 128
godwit, bar-tailed 95
goldcrest 95
goose 11, 59, 73-74, 111, 133, 138, 142
   bar-headed 55
   barnacle 59, 124, 126, *139*
   blue *135*
   brent 143
   Canada 55, *74*
   greylag *75*
   snow *110*, *135*
   white-fronted 120
goshawk 84, 92, 154
ground effect 107, 108-109, *108*, *109*
grouse 120, 124
guillemot 59, 76, 95, 100, 152
gull 11, 38, *39*, 80, 96, 98, 100, 102, 105, 107, 111, 117, 133, 147, 152
   black-headed *134*
   herring 41, 48, 96, *98*, *123*, *132*
   laughing 38
gyrfalcon 126

H

harrier 44, *78*, 79, 92, 96, 125
   Australian spotted *126*
hawk 142
   broad-winged 142
   red-tailed 134, 142, *143*
   sharp-shinned 142, 143, 154
   Swainson's 83
heron 20, 24, 57, 96, 105, 107, 124
   grey 41, 48
hirundines 90, 132
hoatzin 16, *18*, 52
hobby 132, 154
honeyeater 117
hornbill 57, 59, 92, 147
   great pied 90
   ground 53

humerus 28, 30, *30*, *52*, 54
hummingbird 11, 28, 47, *47*, 54, 57, 90, 115-117, 128-129, 139
    amethyst woodstar 115
    broad-tailed *129*
    giant 115
    long-tailed hermit 48, 129, *129*
    ruby-throated 41, 138
    rufous 128
    scintillant *117*
    sicklebill 117
    stripe-tailed *117*

I

Ichthyornis 18
index of manoeuvrability 130

J

jackdaw 69, 96
jay 92
    blue 133
jet stream 55

K

keel 16, 18, *51*, 52
kestrel 11, *35*, *45*, 48, 80, 92, 105, 113-114, *113*, 120, 121, 123, *125*, 126, 147, 149
kingfisher 57, 113
    pied 114
kite *78*, 79, 101, 124, 125
    black 154
    black-shouldered 114
    swallow-tailed 82
kittiwake 80, 123
kiwi 52
knot 8, 133, 144, 147, 150

L

landing 70-76, 109, 152, 154
lee waves 100, 127
lift *15*, 16, 20, 24, *24*, *25*, 26, *26*, 28, 30, 32, *32*, 33, *36*, 37, *37*, 38, *39*, 41, *41*, 43, *43*, 44, *44*, 47, 62, *62*, 65, 70, *73*, 73, 74, 75, *78*, 79, 80, *80*, 82, 102, 109, 113, 117, *117*, 128
    /drag ratio 26, 32, 142
lungs 50, 54, *54*, 55

M

magpie 73, 92
mallard 48, 54, *69*, 92
martin 90, 92, 129-130
    house 129-130, 152
    sand 129-130
maximum range speed *43*, 44, 47, 48, 128, 129, 135, 141, *141*, 142, 147, 149
merlin 66, 126
migration 8, 42, 44, *45*, 47, 48, 55, 59, 95, *95*, 100, 102, *104*, 107, *110*, 111, 120, 121, 128, 135-143, *135*, *138*, *139*, *141*, *143*, 144
minimum power speed 44, 47, 48, 49, 142
mockingbird 41, 133
moorhen *18*, 120
moulting 58-59, *58*, *59*

N

nesting 50, 76, 144, *145*, 147, *147*, 149, *149*, 150
nestling 42, 44, *45*, *47*, 47, *49*, 100, 107, 120, *131*, 144, 147, 149, *149*, 150, 151, 152
nuthatch *77*

O

organs 50, 52, 53

oriole
    black-headed 135
    golden 135
osprey 113, 114, 126, *147*
ostrich 52
owl 31, *31*, 57, 96, 97
    barn *47*, 57
    little 66, *152*
    long-eared 95
    scops *41*, *124*
    short-eared 92, 96, 113, 124
    snowy *37*, *80*, 113, 114
    tawny 92, 95, *95*
oystercatcher 133

P

parrot 52, 57, 97
partridge 120, 124, 132
peafowl 132, *145*
pectoralis muscles 52, *52*, 66, 117
pelican 36, 57, 87, 90, 103, 109, 111
    brown 103, 109
    white 42, 96, *111*
pelvis 15, *17*, 50, *51*
perch *12*, 16, 66, 70, *70*, 76, 83, 95, *123*, 124, 125, *128*, 129, 147, 149
peregrine 48, 74, 124, 125-126, 133, 154
petrel 107, 109, 135
    giant *100*, 123
    storm 115
    Wilson's storm 48, *114*
pheasant *27*, 36, 41, 52, 65, 92, 120, 132
pigeon 11, *17*, 28, 30, *31*, 33, *47*, 52, 53, 55, 62, *62*, 65, 66, 70, 73, *78*, 79, 80, 84, 92, 124, 126, 151, 154
    cape *59*, *118*
    domestic 47
    snow 98
    wood 33, *62*
pitching *78*, *78*, 80
plover, lesser golden 138
power 42, 43, *43*, *45*, 47, *47*, 50, *53*, 65, 90, 96, 97, 111, 117, 127, 141, 142, 150
    induced 43, 44, 47
    parasite 43, 44
    profile 43, 97
power-speed curve 43, 44, 47, 49, 113, 128, 141, *141*, 142
predators, escape from 132-134
preening 31, 55, *55*, 57
Pro-avis 16, 18, 83
ptarmigan *58*
Pterodactylus 14
pterosaurs 14, *14*, 31
puffin 37, 48, *73*, 100, 124
pygostyle *17*, *51*

Q

quail 52, 139
quelea 86
quetzal, resplendent *149*

R

rachis *30*, 31, 36, 58
rail 42, 59, 120
raven 98, 120
razorbill 59, 152
redshank 133, 134
redstart 82, 113, 139
reptile 15, 16, *17*, 28
respiration 53, 54
rhea 52
ribs 15, 16, *17*, 50, 55
robin 49, *49*, *64*, 120
rolling *78*, *78*, 80
rook 20, 21, 76, 120

S

sandpiper 133
    spotted 134
sea breeze 16, 47, 103, *104*
shag 48
    blue-eyed *110*
shearwater 107, 109
    sooty 135
shoulder *17*, 28, 30, 40, 52
shrike, red-backed *123*
side-slip 74, 80, *80*
sinking speed 32, 36, 73, 102
skimmer 108, 109
    black *108*
skua 107, 113, 123, 133, 134
    brown 134
    great *134*
    long-tailed 114, *133*
skylark 66, 132
slope-soaring 98, 100, *100*, 101, 107, 115, 127, 128
slot 36, 37, 43, 65, 90, 102, 120
snipe 144
sparrow, house 31, 48, 54, 55, 97, 113
sparrowhawk 21, 50, 59, *79*, 92, 124-125, 132
speed 28, 32, *32*, 33, *33*, 35, 36, *36*, *43*, 44, 47, *47*, 62, 70, 73, 79, *80*, 90, 92, 96, 124, 129, 141, *141*, 142, *149*, 154
spoonbill
    African 6-7
stall *25*, 26, *26*, 32, *32*, 33, *35*, 36, *36*, 62, 70, *73*, 92, *125*
stamina 50, 53
starling 21, 31, 41, *41*, 42, 48, 55, 70, 73, 86, 92, 96, 113, 120, 133, 149, *149*
    glossy 92
    shining 92
stilt, black-winged *132*
stork 57, 90, 103, 142
    marabou *103*
    painted *70*
    white *12*, 48, *104*
sunbathing 57, *57*
supracoracoideus muscles 52, *52*, 66, 69, 117
swallow 37, 43, 49, 57, 82, 90, 92, 95, 96, 107, 120, 129-130, *131*, 132, 133, 135, 145, 154
    cliff *131*
swan 57, 74-75, 109, 138, 142
    Bewick's 74
    mute 42, 74
    trumpeter *55*, *75*
    whistling *138*
    whooper 74, 83, 142
swift 43, 48, 57, 66, 69, 90, 92, 95, 96, 103, 120, 129-130, 132, 135, 152
    chimney 69
    European 59, 69
    spine-tailed 48
    white-collared 74

T

tail 8, 16, *17*, 19, 31, 37, *37*, *44*, 50, *51*, 79, 80, 82, *82*, 83, *83*, 103, *113*, *126*, 130, 144, 145, 154
taking off 62-66, 69, 90, 102, 103, 120, 128, *128*, 150, 152
tern 37, 82, 107, 113, 123, 133, 147
    Arctic *44*, *45*, 114, *114*, 133
    fairy *149*
    South American *89*
thermal 47, 50, 83, 102, *102*, 103, 127, 128, *128*, 142, 144
    soaring 101-105, 127, 142
thrasher 133
thrush 73, 151
    mistle 133
thrust 16, 24, *24*, 32, *32*, 36, 38, *39*, 41, *41*, 82
tit 41, 70, 95, 97, 147, 151
    blue *11*
    coal *39*

long-tailed 147
toes 15, 16, 28
toucan 52, 92
trachea 54, *54*
traplining 129
triosseal canal 52
tropicbird 123
turbulence 83, 98
turkey 53, 54, 132

V

variable geometry 32-33, 90
vinculum *28*, 30
vulture 11, 36, 50, 57, 73, 90, 101, 102, 103, *103*, 109,
   120, 123, 127-128, *128*, 149
   black 103
   griffon 92, 128, 149-150
   king *128*
   lappet-faced 128
   Rüppell's griffon 55, 128
   turkey *57*

white-backed 102, 128
white-headed 128

W

wagtail 97
warbler 57, 70, 95, 97
   Blackburnian 135
   Cetti's 135
   Dartford 135
   reed *141*
   russet-crowned 135
   sedge 135, 139
   willow 135, 138
weight 24, *24*, 42, 43, 47, 50, 52, 69, 130, 135, 138, 143,
   149, 150, 154
whiffling 73-74, *75*
widowbird 145
   Jackson's 144
   long-tailed *144*
willet 134
willie wagtail 133

wind 47, 74, 77, 96, *100*, 107, 113, 114, 139, 141, 142,
   143, 150
   gradient 105, *106*
wingbeat 28, 31, 55, 62, 65, 74, 92, 108, 113, 114, 117,
   129, 142
   cycle 38, *38*, 39, 40, *40*, *41*, *62*, *115*
   per second 41, 82, 115
wing, evolution of 28
wing-loading 26, 32, 37, 50, 59, 69, 70, 74, 90, 92, 95,
   100, 103, 121, 124, 125, 128, 130, 135, 142, 154
wingspan 26, *27*, 50, 69, 73, 90, 102, 109, 135, 150
wing spreading and folding 30, *30*, 32, 33
wingtip vortices 25-26
woodcock 134
woodpecker 57, 92, 97
   green *39*, 97
wren 90, 92, 135
wren-thrush 120

Y

yawing 78, *78*, 80

ACKNOWLEDGEMENTS

AUTHOR'S ACKNOWLEDGEMENTS

Information on the mechanism and biology of bird flight was gathered from a number of books and many scientific journals. The list of key publications shows my indebtedness to their authors. I am also grateful to the following for advice on aerodynamics and ornithological matters: Andrew Clarke; Bill Hankinson; David Houston; Sinclair Lough; Tom Norcross and Bruce Pearson.

I thank D.R. Waugh for permission to quote from his unpublished thesis 'Predation strategy in aerial feeding birds'; Faber & Faber Ltd for permission to quote from *Swifts* in 'Season's Songs' by Ted Hughes; and A.J. Ward-Smith to use figures in the table on page 92 from *Biophysical Aerodynamics and the Natural Environment*, John Wiley & Sons Inc., 1984.

I am also grateful to Cambridge University Library for supplying a print from *Der Vogelflug als Grundlage der Fliegekunst* by Otto Lilienthal, Gaertners Verlagsbuchhandlung, 1889, used on page 20.

ARTWORK

Eddison Sadd and the author have taken all reasonable steps to obtain permission to use sources of artwork. We thank the authors and publishers of the following books and journals.
17 after Heilmann, G. *The Origin of Birds*, D. Appleton-Company, 1926; 26 Brown, R.H.J. *Biological Reviews*, vol 38, Company of Biologists, 1963; 29 van Tyne, Josselyn and Berger, Andrew *Fundamentals of Ornithology*, John Wiley & Sons, 1959; 39 Brown, R.H.J. *Biological Reviews*, vol 38, Company of Biologists, 1963; 40 after Young, J.Z. *The Life of Vertebrates*, Oxford University Press, 1950; 41 Jenkins, F.A., Dial, K.P. and Goslow, G.E. vol 241, *Science* 1988; 43 Rayner, J.M.V. *Journal of Experimental Biology*, vol 80, Company of Biologists, 1979, Pennycuik, C.J. *Animal Flight*, Edward Arnold, 1972; 47 Pennycuik, C.J. *Journal of Experimental Biology*, vol 49, Company of Biologists, 1968; 54 Freethy, R. *How Birds Work*, Blandford Press, a division of Cassell Publishers, 1982, Bretz, W.L. and Schmidt-Nielsen, K. *Journal of Experimental Biology*, vol 54, Company of Biologists, 1971; 62 Brown, R.H.J. *Journal of Experimental Biology*, vol 25, Company of Biologists, 1948; 100,102,104 Elkins, Norman *Weather and Bird Behaviour*, T. & A.D. Poyser, a division of Academic Press, 1983; 114 after Stolpe, M. and Zimmer, K. *Der Vogelflug*, Akad. Verlag Gesellschaft, 1939, 128 Pennycuik, C.J. vol 111, *Ibis* 1969; 129 Gill, F.B. vol 102, *Auk*, 1985; 141 Pennycuik, C.J. *Animal Flight*, Edward Arnold, 1972; 149 Westerterp, K.R. and Drent, R.H. *Proceedings of the 18th International Ornithological Congress*, vol 1, University of Ottawa Press, 1985.

ARTISTS

Hardlines 24, 25, 26, 27, 32, 35, 36, 39, 41, 43, 47, 54, 62, 78, 80, 93, 97, 100, 102, 104, 106, 115, 127, 128, 129, 141, 149
Sean Milne 15, 17, 28, 29, 30, 31, 40, 51, 52

PHOTOGRAPHERS

t top b below l left r right
Doug Allan/Oxford Scientific Films 114b; Tony Allen/Oxford Scientific Films 153; J.A. Bailey/Ardea, London 84-85; Jack A. Barrie 32, 36, 37, 65, 68, 73r, 74, 75b, 81, 83, 87t, 87b, 136-137, 138, 146; Jen and Des Bartlett/Bruce Coleman Ltd 43, 60-61; Jen and Des Bartlett/Survival Anglia 22-23; Hans and Judy Beste/Ardea, London 126t; Martin Bilfinger 75t; Ennio Boga/Oxford Scientific Films 91, 122; Liz and Tony Bomford 95b, 126b; Liz and Tony Bomford/Ardea, London 84l; Dennis Bright/Swift Picture Library 34-35; Derek Bromhall/Oxford Scientific Films 66-67; Len Bunn/Swift Picture Library 53l; Jane Burton/Bruce Coleman Ltd 18, 56; Bob and Clara Calhoun/Bruce Coleman Ltd 58; Bruce Coleman Ltd 15; Richard Coomber/Planet Earth Pictures 123t; Manfred Danegger 112-113, 125b; Peter Davey/Bruce Coleman Ltd 6-7; P.De Meo/Photo 41r, 45t, 66l, 124; John Downer/Planet Earth Pictures 139; Bacchin Enzo 104, 111, 132; Michael and Patricia Fogden 57t, 116,117, 149t; Jeff Foott/Bruce Coleman Ltd 110t; Jeff Foott/Survival Anglia 130; Geoscience Features Picture Library 14; Francois Gohier/Ardea, London 57b; Su Gooders/Ardea, London 103; C.H. Gomersall/RSPB 134b; F. Greenaway/Bruce Coleman Ltd 46, 148; Martin W. Grosnick/Ardea, London 108; Robert Gross 49b, 92, 123b, 125t; Clem Haagner/Ardea, London 33, 144; Andrew Henley/Biofotos 86; J. Hoogesteger/Biofotos 27, 149b; G.L. Koogman/Oxford Scientific Films 109; Frans Lanting/Bruce Coleman Ltd 152; Michael Leach/Oxford Scientific Films 140; John Mason/Ardea, London 31b; Joe McDonald/Oxford Scientific Films 143; W. Moller/Ardea, London 79; P. Morris/Ardea, London 31t, Musée Marey, Beaune, Cote-d'Or 21; Gertrud Neumann-Denzau 90, 128, 145b; Ben Osborne/Oxford Scientific Films 106-107, 150-151; Stan Osolinski/Oxford Scientific Films 55; Pearson and Prince 82; A. Petretti/Panda Photo 42; Annie Price/Survival Anglia 69, 88-89; Mike Price/Survival Anglia 38; Brian Rogers/Biofotos 78; Science Museum 19; Jonathan Scott/Planet Earth Pictures 53r; Claude Steelman/Survival Anglia 129; Kim Taylor 39, 41l, 63, 77, 97; Kim Taylor/Bruce Coleman Ltd 10-11, 49t, 64, 71, 94, 121, 131; David and Kate Urry/Ardea, London 114t; Paul Ward/Swift Picture Library 110b, 118-119; Kim Westerskov/Natural Images 9, 28, 45b, 59; R. Williams/RSPB 95t; Winfried Wisniewski 12-13, 44, 76, 96r, 96l, 98l, 98-99, 100, 105, 133, 145t; K. Wothe/Bruce Coleman Ltd 147; Günter Ziesler 35tr, 72-73, 101; Günter Ziesler/Bruce Coleman Ltd 2, 70, 134t.

EDDISON SADD

Editor Gilly Abrahams
Designer Nigel Partridge
Artists Sean Milne and Hardlines
Production Controller Claire Kane

Picture Researcher Liz Eddison
Indexer Michael Allaby
Editorial Director Ian Jackson
Creative Director Nick Eddison

598.2185 Burton, Robert, 1941-
BUR
       Birdflight.

$24.95   012944

DATE			
AUG 1 '91			

© THE BAKER & TAYLOR CO.